MATH WORD PROBLEMS
in
1.5 Minutes a Day

D1451527

MATH WORD PROBLEMS
in
15 Minutes a Day

LEARNINGEXPRESS®

NEW YORK

Library of Congress Control Number: 2009927185

A copy of this title is on file with the Library of Congress.

ISBN: 978-1-57685-691-8

Printed in the United States of America

10 9 8 7 6 5 4 3 2

For more information or to place an order, contact LearningExpress at:
  2 Rector Street
  26th Floor
  New York, NY 10006

Or visit us at:
  www.learnatest.com

C O N T E N T S

MATH WORD PROBLEMS

in

15 Minutes a Day

INTRODUCTION

WORD PROBLEMS—PERHAPS two of the most dreaded words in the world of math! Just the mention of them can send people of all ages into a panic.

It is true; word problems present some of the toughest, most challenging problems students face at all levels of mathematics. However, these problems can be the most important because of their real-world applications. These are the problems that we need to solve in our everyday lives, problems about paying bills, taking out loans, saving money, interpreting data; the list goes on. These are the questions we cannot ignore or avoid, but they are also the problems we would love to live without! Nonetheless, we need to face our fears and rise up to the challenge. With a little help, word problems can be tackled just like other obstacles we face.

We can make these problems much more manageable by using the strategies discussed in this book. Taking time to read the question carefully, underlining key words and concepts, breaking apart large problems into smaller pieces, and checking your answer to make sure it is reasonable are a few of the ways to face up to the problems at hand. Many other strategies will be explained in detail, such as looking for a pattern, drawing a picture, using logical reasoning, working backward, making an organized list, plus many others. You'll become skilled in solving math word problems by building a long list of strategies in your problem-solving repertoire.

USING YOUR BOOK

By design, this book breaks down complicated word problems into manageable lessons that pinpoint particular topics. Since each lesson can be completed in about 15 minutes, improving your word-problem skills is only moments away.

THE BOOK AT A GLANCE

Exactly what will you find in this book? It begins with the introduction, outlining the key concepts and tasks presented. Following the introduction is a pretest of 30 questions. This pretest will give a preview of the topics presented in each of the 30 lessons in the same order they are taught in the book. Taking this pretest will give you information about topics with which you may already be comfortable, along with areas in which you may be hoping to improve. Next, each of the 30 lessons breaks down the topics by giving examples and then provides practice in each area. Answer explanations are also included to help you in your quest to be skilled at solving word problems. Following the lessons is a posttest. Designed with the same format as the pretest, this will tell you areas of improvement and strengths you have acquired along the way.

The main part of the book is divided into five sections:

1. Solving Word Problems—Selecting the Right Strategy
2. Number Sense Word Problems—Making Them Make Sense
3. Algebra Word Problems—Finding the Unknown in Unknown Territory
4. Geometry and Measurement Word Problems—Learn How to Measure Up
5. Probability and Statistics Word Problems—Increase Your Chances

Each section contains a series of lessons focused on the section topic. Each lesson provides details about a particular type of math word problem, accompanied by practice problems to sharpen your skills. In addition, there are a number of helpful tips and shortcuts pointed out for each lesson area.

By using the examples, tips, and practice, you can improve your math word-problem skills.

The days of dreaded math word problems are almost over. Face the challenge of math word problems by using the skills and strategies presented in this book, and watch your progress increase. In just 15 minutes a day, you can master word problems in most every area of mathematics and prove to yourself at last, there is no need to panic!

P R E T E S T

THIS PRETEST HAS 30 multiple-choice questions, each corresponding to the topics covered in the 30 lessons of this book. Use this pretest to find out what you already know about solving math word problems, and to identify areas where there is room for improvement. Read and answer each problem carefully by selecting the best answer choice. Do not be afraid to show your work; it is the cornerstone of problem solving and will also help you to identify any mistakes.

When you have finished the pretest, check your answers with the answer key beginning on page 12. Do not worry if you did not get every question correct; solving math word problems can be tricky, and this is why you are using this book in the first place. Do note, however, the question numbers of any problems you did answer incorrectly. These problem numbers match the lessons containing the skills tested in the question. Spend extra time on those lessons as you go through this book in order to see your improvement of those skills.

PRETEST

1. Which statement means the same as "three less than a number, n"?
   a. $3 - n$
   b. $3 \div n$
   c. $n - 3$
   d. $n < 3$

2. Charlie has two choices of a lunch and three choices of a drink. How many different ways can he choose a lunch and a drink?
   a. 3
   b. 5
   c. 6
   d. 8

3. Gerry left his house and headed for the store with some money in his pocket. First, he spent $\frac{1}{2}$ of his money on new trading cards. Then, he spent $1.00 on an ice cream cone. Finally, he spent $2.00 on a drink. After this trip, he had $7.00 left over. How much money did Gerry have before heading to the store?
   a. $7.00
   b. $20.00
   c. $24.00
   d. $40.00

4. There are 60 students in the school band and 70 students in the chorus. If there are 20 students in both band and chorus, how many students are in the chorus only?
   a. 10 students
   b. 20 students
   c. 40 students
   d. 50 students

5. There are 14 people who work in an office. If each person sends each other person at the office one e-mail on a particular day and there is only one e-mail between every possible pair, how many e-mails will be sent that day?

   a. 14

   b. 84

   c. 105

   d. 182

6. On Saturday, the low temperature was 9°. On Sunday, the low temperature was 14° below the low temperature on Saturday. What was the low temperature on Sunday?

   a. −14

   b. −5

   c. 14

   d. 25

7. A recipe that Carl is using calls for $\frac{3}{4}$ cup of sugar for each dozen cookies he makes. How many cups of sugar will he need to make three dozen cookies?

   a. 2 cups

   b. $2\frac{1}{4}$ cups

   c. $2\frac{1}{2}$ cups

   d. 3 cups

8. Six adults and five students are going to a show. If adult tickets cost $7.50 each and student tickets cost $5.25 each, how much money will the group spend on tickets?

   a. $12.75

   b. $63.75

   c. $71.25

   d. $76.50

9. If a car moving at a constant rate travels 385 miles in 7 hours, what is the rate of the car in miles per hour?

   a. 55 miles per hour

   b. 56 miles per hour

   c. 57 miles per hour

   d. 62 miles per hour

10. If 8 gallons of gasoline cost $25.92, what is the cost of 15 gallons?

    a. $3.24

    b. $32.40

    c. $38.88

    d. $48.60

11. What is 40% of 50?

    a. 20

    b. 40

    c. 60

    d. 125

12. Jason is a salesperson and earns 4% commission on his total sales for each month. If he made a total of $2,450 in sales this month, what is the amount of commission he earned?

    a. $40

    b. $61.50

    c. $98

    d. $980

13. The sum of 5 and two times a number is equal to 27. What is the number?

    a. 11

    b. 13.5

    c. 19

    d. 22

14. There are 193 students going on a school trip. If each bus can fit 51 students, what is the least number of buses they will need for the trip?

    a. 2

    b. 3

    c. 4

    d. 5

15. The speed of light can be expressed as 6.71×10^8 miles per hour in scientific notation. What is this number in standard notation?

    a. 0.000000671

    b. 0.0000000671

    c. 67,100,000

    d. 671,000,000

16. Janice mixes candy that costs $2.00 per pound with candy that costs $4.00 per pound. She buys twice as much of the $2.00 per pound candy than she buys of the $4.00 per pound candy. If she spent a total of $32.00, how many pounds of the $2.00 per pound candy did she buy?

   a. 2 pounds

   b. 4 pounds

   c. 6 pounds

   d. 8 pounds

17. Two angles are complementary. If the measure of one angle is 10 more than the measure of the other, what is the measure of the larger angle?

   a. 10°

   b. 40°

   c. 50°

   d. 90°

18. The measure of the second angle of a triangle is 20° more than the measure of the first angle of the triangle. The measure of the third angle of the triangle is 40° more than the measure of the first angle. What is the measure of the largest angle of this triangle?

   a. 20°

   b. 40°

   c. 60°

   d. 80°

19. The measure of angle A in parallelogram $ABCD$ is 85°. What is the measure of angle D?

   a. 85°

   b. 95°

   c. 265°

   d. 275°

20. Two triangles are similar. The measures of the shortest sides of each triangle are 3 units and 6 units, respectively. If the measure of the longest side of the small triangle is 12, what is the measure of the longest side of the larger triangle?

   a. 6

   b. 18

   c. 24

   d. 30

21. The length of a rectangular picture frame is 5 inches more than twice the width. If the perimeter of the frame is 70 inches, what is the length of the frame?

   a. 10 inches

   b. 25 inches

   c. 50 inches

   d. 60 inches

22. The radius of a circle is 5 cm. What is the area in square units of the circle in terms of π?

   a. 5π

   b. 10π

   c. 20π

   d. 25π

23. Curtis wants to paint the ceiling of a room that has a length of 16 feet and a width of 10 feet. If one can of paint will cover about 75 square feet, what is the minimum number of cans that he needs?

   a. 1

   b. 2

   c. 3

   d. 4

24. Hope is covering a can in the shape of a cylinder with wrapping paper. What is the minimum amount of paper, to the nearest tenth of a square inch, that she will need if the can has a height of 4 inches and the base has a radius of 2 inches? Use 3.14159 for π and the formula $SA = 2\pi r^2 + \pi dh$.

   a. 25.1 square inches

   b. 50.3 square inches

   c. 75.4 square inches

   d. 89.7 square inches

25. Katie is filling a container in the shape of a rectangular prism with popcorn. What is the volume of the container if the width is 3 inches, the length is 4 inches, and the height is 2.5 inches?

   a. 9.5 cubic inches

   b. 12 cubic inches

   c. 30 cubic inches

   d. 59 cubic inches

26. Jackson plots the two points (2,4) and (–3,5) on a coordinate plane. What is the slope of the line that would connect these two points?

   a. $-\frac{1}{5}$

   b. $\frac{1}{5}$

   c. –5

   d. 5

27. A spinner has 9 equal sections numbered 1–9. What is the probability of getting an even number with one spin?

   a. $\frac{1}{9}$

   b. $\frac{2}{9}$

   c. $\frac{4}{9}$

   d. $\frac{1}{2}$

28. Michael selects one letter from the word ALGEBRA. What is the probability that he selects a G or a vowel?

   a. $\frac{1}{7}$

   b. $\frac{2}{7}$

   c. $\frac{3}{7}$

   d. $\frac{4}{7}$

29. What is the total number of ways that five different books can be arranged on a shelf?

   a. 5

   b. 24

   c. 25

   d. 120

30. The heights of five people from Cheryl's class are 60 inches, 58 inches, 70 inches, 60 inches, and 63 inches. What height is the mode of this data?

   a. 58 inches

   b. 60 inches

   c. 62.2 inches

   d. 70 inches

ANSWERS

1. c. The key phrase *less than* means "subtract from a number." Thus, three *less than* a number, *n*, is represented by the expression $n - 3$.

2. c. One strategy to use to solve this problem is to make an organized list. Since there are two choices of lunch, call one lunch A and the other lunch B. Because there are three choices of drinks, represent each as drink C, drink D, and drink E. Make a list of all possible choices containing exactly one lunch choice and one drink choice.

Lunch A with drink C   Lunch B with drink C

Lunch A with drink D   Lunch B with drink D

Lunch A with drink E   Lunch B with drink E

There are six different combinations in the list, so there are six possibilities.

3. b. Use the strategy of working backward to solve this problem. Begin with the $7.00 he had left over at the end of the trip to the store. Then, do the opposite to work backward to find how much he had at the start. Add the $2.00 he spent on a drink to get $9.00, and add another $1.00 for the ice cream cone to get $10.00. At this point, he had spent half of his money on trading cards, so double $10.00 to get the original amount of $20.00.

4. d. Draw a Venn diagram to help with this question. The diagram should contain two circles that overlap, such as the following diagram. Label one circle *band* and the other *chorus*.

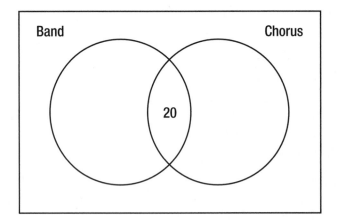

The 20 students in both band and chorus should be placed in the section where the two circles overlap. Since there are 60 total students in the band, this includes the 20 students in both groups. Therefore, there are 40 students in band only. In the same way, the 70 students in chorus represents the students in chorus only and in both groups, so there are 70 – 20 = 50 in chorus only. The completed diagram is shown next.

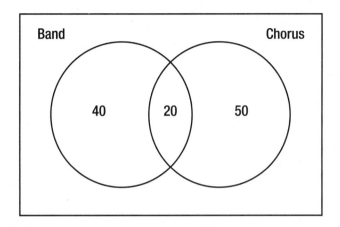

5. c. Use the strategy of solving a simpler problem by using smaller numbers of people in the office. If there were three people in the office, the first person would send three e-mails, the second would send two e-mails, and the third only one. Remember that there is only one e-mail between each pair of people. Thus, the pattern for three people is 3 + 2 + 1 = 6 e-mails. Therefore, the pattern for 14 people is 14 + 13 + 12 + 11 + 10 + 9 + 8 + 7 + 6 + 5 + 4 + 3 + 2 + 1 = 105 e-mails.

6. b. The low temperature on Sunday was 14° below the low temperature on Saturday. To solve this problem, subtract: 9 – 14 = –5°.

7. b. To solve this problem, multiply $\frac{3}{4}$ cup of sugar by 3 for each dozen. $\frac{3}{4} \times \frac{3}{1} = \frac{9}{4} = 2\frac{1}{4}$ cups of sugar.

8. c. Multiply 6 by the cost of an adult ticket: 6 × $7.50 = $45.00.
Multiply 5 by the cost of a student ticket: 5 × $5.25 = $26.25.
Add these two amounts together to get the total spent on the tickets: $45.00 + $26.25 = $71.25.

9. a. Divide the total miles by the total time traveled to find the rate.
$\frac{385 \text{ miles}}{7 \text{ hours}}$ = 55 miles per hour.

10. d. Set up the proportion:
$\frac{\$25.92}{8 \text{ gallons}} = \frac{\$x}{15 \text{ gallons}}$
Be sure to line up the units. Cross multiply to get 8x = 388.8. Divide each side of the equation by 8:
$\frac{8x}{8} = \frac{388.8}{8}$
x = $48.60

11. a. The key word *of* tells you to multiply 40% by 50. Change the percent to a decimal before multiplying: 0.40 × 50 = 20.

12. c. Find 4% of $2,450 to find the total commission. Change the percent to a decimal before multiplying: $0.04 \times \$2,450 = \$98.00$.

13. a. Translate the statement from words into mathematical symbols. Let $n =$ a number. *Sum* is a key word for addition and *two times a number* can be expressed as $2n$. The equation is $5 + 2n = 27$. Subtract 5 from each side of the equation to get $2n = 22$. Divide each side of the equation by 2.

$$\frac{2n}{2} = \frac{22}{2}$$

$$n = 11$$

14. c. Let $n =$ the least number of buses needed. Since each bus can fit 51 students, the inequality is $51n \leq 193$. Divide each side of the inequality by 51:

$$\frac{51n}{51} \leq \frac{193}{51}$$

$$n = 3.7843$$

This amount needs to be rounded, so there is enough room for all students. The least number of buses is 4.

15. d. To convert from scientific notation to standard notation, find the exponent on the base of 10. This exponent tells you how many places to move the decimal point; count to the right if the exponent is positive and to the left if it is negative. Start at the decimal point in 6.71 and count to the right eight places. Add zeros where necessary. The number becomes 671,000,000.

16. d. Let $x =$ the number of pounds of the $4.00 candy, and let $2x =$ the number of pounds of the $2.00 candy. Write the equation:

$$\$2.00(2x) + \$4.00x = \$32.00$$

Multiply within the equation to get

$$4x + 4x = 32$$

Combine like terms to get

$$8x = 32$$

Divide each side of the equation by 8:

$$\frac{8x}{8} = \frac{32}{8}$$

Thus

$$x = 4$$

$$2x = 2(4)$$

$$= 8$$

Janice bought 4 pounds of the $4.00 candy and 8 pounds of the $2.00 candy.

17. c. Let x = the smaller angle and let $x + 10$ = the larger angle. The term *complementary* means that the sum of the two angles is 90°. To write the equation, add the two angles and set the sum equal to 90:

$x + x + 10 = 90$

Combine like terms to get

$2x + 10 = 90$

Subtract 10 from each side of the equation: $2x + 10 - 10 = 90 - 10$
The equation simplifies to

$2x = 80$

Divide each side of the equation by 2:

$\frac{2x}{2} = \frac{80}{2}$

$x = 40$

Therefore, the smaller angle is 40, and the larger angle is $x + 10 = 40 + 10 = 50°$.

18. d. Let x = the measure of the first angle. Let $x + 20$ = the measure of the second angle. Let $x + 40$ = the measure of the third angle. The sum of the three angles of a triangle is equal to 180°, so add the three expressions and set the sum equal to 180:

$x + x + 20 + x + 40 = 180$

Combine like terms:

$3x + 60 = 180$

Subtract 60 from each side of the equation:

$3x + 60 - 60 = 180 - 60$

$3x = 120$

Divide each side of the equation by 3:

$\frac{3x}{3} = \frac{120}{3}$

$x = 40$

Therefore, the first angle is 40°, the second angle is 40 + 20 = 60°, and the third angle is 40 + 40 = 80°. The largest angle is 80°.

19. b. Angle A and angle D in parallelogram $ABCD$ are consecutive angles, or angles that are next to each other. Therefore, the sum of their measures is 180°. If the measure of angle A is 85°, subtract 180 − 85 = 95 to find the measure of angle D. Angle D is 95°.

20. c. Use the given sides of both similar triangles to form a proportion. Let x = the longest side of the large triangle. Be sure to line up corresponding labels. The proportion could be set up as:

$$\frac{\text{shortest side of small triangle}}{\text{longest side of small triangle}} = \frac{\text{shortest side of large triangle}}{\text{longest side of large triangle}}$$

Substitute the given values:

$$\frac{3}{12} = \frac{6}{x}$$

Cross multiply to get $3x = 72$. Divide each side of the equation by 3:

$$\frac{3x}{3} = \frac{72}{3}$$
$$x = 24$$

The longest side of the larger triangle is 24 units.

21. b. Let x = the measure of the width, and let $2x + 5$ = the measure of the length. Use the perimeter formula $P = 2w + 2l$. Substitute the *let* statements and the fact that the perimeter $P = 70$ to get the equation

$$70 = 2x + 2(2x + 5)$$

Use the distributive property to get:

$$70 = 2x + 4x + 10$$

Combine like terms:

$$70 = 6x + 10$$

Subtract 10 from each side of the equation:

$$70 - 10 = 6x + 10 - 10$$

The equation simplifies to:

$$60 = 6x$$

Divide each side of the equation by 6:

$$\frac{60}{6} = \frac{6x}{6}$$
$$10 = x$$

Use the expression for the length $(2x + 5)$ and substitute $x = 10$ to find the measure of the length:

$$2(10) + 5 = 20 + 5 = 25$$

The length is 25 inches.

22. d. The area of the circle can be found by using the formula $A = \pi r^2$, where r is the radius of the circle. Because the question states that the answer is in terms of π, leave pi in your answer. The radius is 5, so substitute 5 into the formula: $A = \pi(5)^2 = 25\pi$ square units.

23. c. First, find the area of the ceiling using the formula $A = l \times w$. Since the length is 16 feet and the width is 10 feet, the area is $16 \times 10 = 160$ square feet. One can of paint is needed for every 75 square feet, so divide the total area by 75:

$$160 \div 75 = 2.133$$

Because this amount is more than 2, he will need to buy 3 cans of paint to cover the ceiling.

24. c. Find the surface area of the cylinder by using the formula $SA = 2\pi r^2$, where r is the radius of the base, d is the diameter of the base, and h is the height of the cylinder. From the information in the problem, $r = 2$, $d = 4$ (double the radius), and $h = 4$. Substitute the given values into the formula and use $\pi = 3.14159$:

$$SA = 2(3.14159)(2)^2 + (3.14159)(4)(4)$$

Evaluate the exponent:

$$SA = 2(3.14159)(4) + (3.14159)(4)(4)$$

Multiply in each term: $SA = 25.13272 + 50.26544$

Add to find the surface area:

$$SA = 75.39816$$

Round this value to the nearest tenth:

$$SA = 75.4 \text{ square inches}$$

25. c. Use the formula $volume = length \times width \times height$ ($V = l \times w \times h$). Substitute the given values into the formula:

$$V = 4 \times 3 \times 2.5$$
$$= 30 \text{ cubic inches}$$

26. a. Use the slope formula:

$$m = \frac{\text{change in } y}{\text{change in } x} = \frac{y_1 - y_2}{x_1 - x_2}$$

Substitute the coordinates of the points. Use (2,4) for (x_1,y_1) and (–3,5) for (x_2,y_2). The formula becomes:

$$m = \frac{4-5}{2-(-3)} = \frac{-1}{2+3} = \frac{-1}{5}$$

27. c. The probability of an event E is equal to P(E) =

$$P(E) = \frac{\text{number of ways event can occur}}{\text{total number of possible outcomes}}$$

There are four even-numbered sections {2, 4, 6, 8}, so the number of ways the event can occur is 4. There are a total of 9 sections, so the total number of possible outcomes is 9. The probability is $\frac{4}{9}$.

28. d. The probability of an event E is equal to

$$P(E) = \frac{\text{number of ways event can occur}}{\text{total number of possible outcomes}}$$

There are 7 letters in the word ALGEBRA, so the total number of possible outcomes is 7. There is one G and three vowels in the word. Thus, the probability of selecting a G is $\frac{1}{7}$ and the probability of selecting a vowel is $\frac{3}{7}$. The probability of selecting a G or a vowel becomes $\frac{1}{7} + \frac{3}{7} = \frac{4}{7}$.

29. d. The total number of ways is equal to the number of permutations of five things taken five at a time. This is equal to $5 \times 4 \times 3 \times 2 \times 1 = 120$ different orders.

30. b. The mode of a set of data is the value that occurs the most often. The value that appears the most in the list of data is 60 inches.

ⓈⒺⒸⓉⒾⓄⓃ 1

solving word problems—selecting the right strategy

MATH WORD PROBLEMS can be both confusing and intimidating. However, by having skills and strategies in your back pocket, you can gain confidence and be successful at solving them.

This section details the steps to take while you are solving word problems, as well as a variety of approaches for finding a solution. Since most every word problem appears differently at first glance, similarities and differences in certain problem types will be highlighted to help you pinpoint the best approach to each problem. Not everyone solves problems the same way; every brain works a little differently. What appears obvious to one person may not be at all clear to another. Keep in mind, however, these strategies are suggestions of how to approach a certain problem type, but there are numerous ways to solve most math word problems. You may find an easier approach yourself, but you can use the detailed explanations to help you find a method of solution or an alternative way of solving the problem.

This section introduces you to basic word-problem strategies, including:

- word-problem solving steps
- using key words
- looking for a pattern
- making an organized list
- working backward

- using logical reasoning
- making a table
- drawing a picture
- solving a simpler problem
- guess and check

the steps to solving word problems and valuable key words

If two wrongs don't make a right, try three.

—AUTHOR UNKNOWN

One of the best ways to solve a complicated problem is to break it down into smaller, more manageable pieces. This chapter details the steps to solving math word problems and a four-step method, along with a discussion of common key words and phrases used in math word problems.

Apply the strategies from this lesson throughout the book and you will be on your way to becoming a successful word-problem solver.

THE STEPS TO SOLVING WORD PROBLEMS

Solving math word problems is a daunting task for some people. However, by having a game plan in mind, even the most difficult problems can be solved.

1. The first step is to *read and understand the question*. For each problem, be sure to carefully read the text the first time through to get the general picture of what the question is asking. At this point, list the information given and summarize the problem in your own words.

2. The second step is to *make a plan* of attack. This will be your approach to solve the problem. In this step, underline any key words, numbers, or phrases you see in the question. This step will assist you in determining the correct operation or operations that should be used. In addition, cross out any extra information that is not necessary to solve the problem.

3. The third step is to *carry out the plan*. In this step, use the plan outlined in step two. This may include strategies such as drawing a picture or diagram, extending a pattern, writing an equation, using a formula, or using guess and check. Each of these strategies will be explained in the various lessons throughout this book. Identify the plan to solve the problem.

4. For the final step, *check your answer* to be sure it is reasonable. Many times, an answer may be the result of an error in setting up the problem, and you may not realize it if the solution is not checked. Review the work done for the problem and the answer reached: does it make sense? If at all possible, check your work in a different way from the way you used to solve the problem. For example, after you have solved an equation, check your work by substituting in the answer and using order of operations to check, instead of simply solving the equation a second time.

....

TIP: The **problem solving steps** can be summarized by the following:

1. Read and understand the question.
2. Make a plan.
3. Carry out the plan.
4. Check your work.

....

To get started, let's practice these steps using an example.

Example

Charlie brought in 3 boxes of cookies to share with his class of 22 students. Chocolate chip cookies are his favorite kind. If there are 18 cookies in each box and each student needs to get the same number of cookies, how many cookies will be left over?

1. *Read and understand the question.* This question asks for the number of cookies left over after they are divided up among students in a class.
2. *Make up a plan.* First, figure out the total number of cookies. Multiply the number of boxes by the number of cookies in each box. Then, divide the total by the number of students in the class. The remainder will represent the number of cookies left over. The question is shown below with important information underlined, and extra information crossed out.

    Charlie brought in *3 boxes* of cookies to share with his class of *22 students*. If there are *18 cookies in each box* and each student needs to get the same number of cookies, *how many cookies will be left over*?

3. *Carry out the plan.* Multiply the number of boxes by the number of cookies in each box to find the total:

    $18 \times 3 = 54$

    Then, divide 54 by the total number of students in the class:

    $54 \div 22 = 2$

    with a remainder of 10. This gives each student 2 cookies, with 10 left over. The solution is that there will be 10 cookies left over.

4. *Check your work.* If each student gets 2 cookies, this is a total of $22 \times 2 = 44$ cookies to be eaten. Since there were 10 left over, the total number of cookies is $44 + 10 = 54$. Because this value is the same as the total number of cookies in the three boxes, the answer is reasonable.

These important steps will be modeled and applied throughout this book to help you solve problems. Use this procedure as a way to tackle any math word problem you may meet on your road to word problem success. Begin your journey by working through the practice set below.

PRACTICE 1

Put the problem-solving steps in the correct order from 1 to 4. Place the number of the step on the line.

_____ Carry out the plan.

_____ Make a plan.

_____ Check your work.

_____ Read and understand the question.

USING KEY WORDS

There are many common key words that appear in math word problems, and using key words and phrases is very helpful when you are deciding on the operation or operations needed. These key words should be underlined or highlighted when you are devising your plan to solve each problem. Here are examples of some frequently used key words and phrases, along with the symbol or operation they represent.

Addition (+)
sum, increased, combine, plus, more than, total

Subtraction (–)
difference, decreased, less than, take away

Multiplication (×)
product, times, factor, twice (×2), triple (×3)

Division (÷)
quotient, divide, into, out of, split up, break up

Equal (=)
is, total, result, same as, equivalent to

Greater than (>)
is more than, is greater than, is larger than, above

Greater than or equal to (≥)

minimum, at least, is not less than, is not smaller than

Less than (<)

is smaller than, is less than, below

Less than or equal to (≤)

maximum, at most, is not more than, is not greater than

The key words mentioned in the list are often used in math word problems, so look out for these in the problems in this book, or any other word problems you may come across.

....

TIP: Be careful with certain key phrases that are similar.

more than can mean "addition"
is more than can mean "is greater than (>)"
less than can mean "subtract"
is less than can mean "is less than (<)"

....

PRACTICE 2

Match the terms below to the correct symbol:

at least	+
at most	−
sum	×
product	÷
is the result	≤
quotient	≥
difference	=

TRANSLATING FROM WORDS TO MATH SYMBOLS

Translating from words into symbols is very much like converting from one language to another. Look for the important vocabulary and use the numbers mentioned in the question to help you write a number sentence. In most cases, the order in which the values and key words are used in the statement is the same as the order in which they will appear in the numerical sentence.

Translate the following examples from words into mathematical symbols:

1. Five increased by 10
   The key phrase *increased by* means to "add 5 and 10," so the expression is $5 + 10$.

2. The product of 6 and 8
   The key word *product* means "to multiply," so the expression is 6×8.

3. The quotient of 70 and 7
   The key word *quotient* means "to divide," so the expression is $70 \div 7$.

4. The difference of 9 and 4 is the same as 5.
   The key word *difference* means "subtraction," and the phrase *is the same as* means "is equal to." The equation is $9 - 4 = 5$.

5. Five is not greater than 7.
   The key phrase *is not greater than* means "is less than or equal to," so the inequality is $5 \leq 7$.

6. The sum of 6 and 3 is not less than 8.
   The key word *sum* means "add 6 and 3," and the key phrase *is not less than* means "is greater than or equal to." The inequality becomes $6 + 3 \geq 8$.

7. Three less than 11 is equal to 8.
   The key phrase *less than* means subtraction from a quantity, and the phrase *is equal to* means "equals." The equation is $11 - 3 = 8$.

....

TIP: The order of the numbers and symbols does not always stay the same for some key words and phrases. In a few cases, the order is reversed.
   For example, 6 *less than* 10 or 6 *subtracted from* 10 both translate to $10 - 6$.

....

These examples represent a sample of some of the important vocabulary used in many math word problems. Becoming familiar with these words and phrases is a good way to improve your word-problem solving skills. Work on the following practice problems to see how many of the key words and phrases you have learned.

PRACTICE 3

Translate each statement from words into math symbols.

1. Five more than 8.

2. The difference between 10 and 4.

3. Twice 40 is subtracted from 100.

4. The quotient of 21 and 7 is not more than 4.

5. Twenty decreased by 6 has a result of 14.

6. The product of 12 and 3 is at least 36.

ANSWERS

Practice 1

3	Carry out the plan.
2	Make up a plan.
4	Check your work.
1	Read and understand the question.

Practice 2

at least	$\geq$
at most	$\leq$
sum	$+$
product	$\times$
is the result	$=$
quotient	$\div$
difference	$-$

Practice 3

1. $8 + 5$. The key phrase *more than* means "addition."
2. $10 - 4$. The key word *difference* means "subtraction."
3. $100 - 2 \times 40$. The phrase *twice forty* means "multiply 40 by 2." This value is *subtracted from* 100, so the order in the expression is the reverse of the order in the statement.
4. $21 \div 7 \leq 4$. The key word *quotient* means "divide." The key phrase *is not more than* means "is less than or equal to."
5. $20 - 6 = 14$. The key phrase *decreased by* means "subtraction" and the key phrase *has a result of* means "is equal."
6. $12 \times 3 \geq 36$. The key word *product* means "multiply" and the key phrase *at least* means "greater than or equal to."

the strategies of looking for a pattern and making an organized list

Mathematics is written for mathematicians.
—NICHOLAUS COPERNICUS (1473–1543)

Having a plan in mind is an important step in solving math word problems. The two strategies presented in this section will show you how to look for patterns and make organized lists.

LOOKING FOR A PATTERN

Some math problems can appear very complicated when you are first reading the question. However, with many problems, a pattern or a trend may emerge to make solving the problem easier. Often, you can extend the pattern, so finding out what comes next in a series of numbers or the pattern may simplify what is actually happening in a problem. Looking for a pattern is one key strategy that will be discussed in this section. You will also find examples of problems that can be solved by this method.

One type in this category is number pattern problems. In this type of question, a series of numbers is given, and you may be asked to find the next number in the sequence. For example, take a look at the following pattern.

2, 4, 6, 8, 10, _____

What is the next number in the list?

This is a list of even numbers, or numbers beginning with the number 2, with 2 added each time to find the next number. The next number in the list would be the next even number, 12, which is also equal to $10 + 2$.

The pattern may also be decreasing, as in the following list.

100, 80, 60, 40, _____

In this list, 20 is subtracted from the previous term to get the next term. The next number in the list would be $40 - 20 = 20$.

....

TIP: If a pattern appears to be increasing, check first to see if addition or multiplication is the rule. If the pattern is decreasing, check first to see if subtraction or division is the rule.

....

Number patterns can also be found in a context or situation, which is the case in the following question.

Example

The end of the regular school day is 2 P.M. at Hanford High School. There are 1,200 students at the school. During the first hour after school, half of the students in the building leave and the other half remain for extra help and various after school activities. During the second hour after the school day, half of the remaining students leave. If this pattern continues, how many students are left in the building four hours after the end of the school day?

To solve this problem, use the steps to solving word problems explained in Lesson 1.

Read and understand the question. The number of students in a school decreases each hour after the school day ends. The question asks for the number of students in the building four hours after the school day ended.

Make a plan. Start with the total number of students in the building and make a decreasing pattern. Continue the pattern until the number of students remaining four hours after school is reached.

Carry out the plan. The number of students in the building at the end of the school day is 1,200. Thus, after one hour only half remain, or 600 students. After the second hour, half of 600 remain, or 300 students. After the third hour, half of 300 remain, or 150 students. Finally, after the fourth hour, half of 150 remain, or 75 students. In summary, the pattern is: 1,200, 600, 300, 150, 75. There are 75 students left in the building four hours after the end of the regular school day.

Check your work. Start with the 75 students remaining after four hours and work back to the end of the school day: $75 \times 2 = 150$ after 3 hours; $150 \times 2 = 300$ after 2 hours; $300 \times 2 = 600$ after 1 hour; and $600 \times 2 = 1,200$ at the end of the school day.

Patterns can also be displayed in table format. The table may be horizontal or vertical, but both are solved in the same manner. Take the following table.

What is the missing value in the table?

x	y
1	3
2	6
3	9
4	12
5	?

Use the steps to solving word problems to work through the problem.

Read and understand the question. You are looking for the missing value in a table where most of the numbers are given.

Make a plan. Look for a pattern in the table to help you figure out the missing number. This pattern can be a horizontal (across) or vertical (up/down) pattern.

Carry out the plan. In this table, each of the values is filled in except for the location of the question mark, so look for a pattern with the other numbers. For each given x-value, the corresponding y-value appears to be three times the x-value. Be sure to check each row to make sure the pattern holds true for all numbers in the table. By using this pattern, the missing number is $3 \times 5 = 15$.

Check your answer. Another way to view this table is vertically. As the y-values increase, each number is three more than the previous one. By using this method, the missing number is $12 + 3 = 15$, which is the same answer found by the other method.

In each of the questions in the practice that follows, apply the strategy of looking for a pattern to help simplify the problem. Use the answer explanations at the end of the lesson to check your work and your solution.

PRACTICE 1

1. Jake made a pattern by writing the following numbers.

2, 4, 8, 16, 32, _____

If he continued the pattern, what would be the next number in this list?

2. A pattern of numbers is as follows: 1, 3, 7, 15, 31, . . . What would be the eighth number in the pattern?

3. There are four people at a meeting. If each person will shake hands with another person in the room exactly once, how many handshakes will take place at the meeting?

4. A function table is shown next.

x	y
1	3
2	5
3	7
4	9
5	?

What is the unknown number in the table?

MAKING AN ORGANIZED LIST

Another strategy that can be used to solve many math word problems is making an organized list. This strategy is similar to looking for a pattern, but this time, you will make a list that will include all possibilities in a situation. Take a look at the problem below.

Example

Richard has 3 different ties: blue, purple, and red, and four different dress shirts: green, yellow, white, and black. How many shirt-and-tie combinations can be made by selecting 1 shirt and 1 tie?

Use the problem-solving process to solve this question, along with the strategy of making an organized list.

Read and understand the question. The question is asking for the total number of possibilities when selecting 1 shirt and 1 tie. There are 3 ties and 4 shirts from which to choose.

Make up a plan. Make a list of all of the combinations of 1 shirt with 1 tie. Then count the total number of possibilities to find the answer.

Carry out the plan. Make an organized list pairing each tie with each shirt. An organized list could appear as follows:

blue—green	purple—green	red—green
blue—yellow	purple—yellow	red—yellow
blue—white	purple—white	red—white
blue—black	purple—black	red—black

This is a total of 12 different pairs; therefore, there are 12 different possibilities.

Check your work. Since there are 3 different ties being paired with four different shirts, there should be 4 + 4 + 4 =12 possibilities. This is the same as the number of items in the organized list.

....

TIP: When making an organized list, you can also use abbreviations for the words in your list. For example, b—g would represent the blue tie with the green shirt. This may make it easier and faster to construct your list.

....

Try the following practice set. Use the strategy of making an organized list to help find a reasonable answer.

PRACTICE 2

1. At an ice cream shop, patrons can select chocolate, strawberry, or vanilla
ice cream. They can also select if they want the ice cream in a cone or a
dish. If only one flavor can be selected, how many different types of ice
cream desserts can be chosen?

2. Janice can select History, Math, or English for her period 1 class. For her
period 2 class, she can select Art, Music, Health, or Science. If she selects
exactly one class for each of these periods, how many different combinations of classes are there?

3. At the Bistro Burger restaurant, a customer can select a plain or sesame
roll for their burger. They can also select one of four toppings: ketchup,
mustard, mayonnaise, or relish. If a customer selects exactly one type of
roll and one topping, how many different burgers can be made?

ANSWERS

Practice 1

1. *Read and understand the question.* The question is looking for the next number in a pattern.
Make a plan. Find the pattern between the given numbers in the list, and
apply this rule to the last known value in the list.
Carry out the plan. The values in the list are doubling; each number is
equal to the previous number multiplied by 2. To find the next number,
multiply 32 by 2 to get 64. The next number in the list is 64.
Check your work. One way to check this solution is to work from the last
number to the first by doing the opposite, or inverse operation. Start with
64 and divide by 2 to get the previous term of 32. Then continue this
process to be sure all numbers were generated the same way: $32 \div 2 = 16$;
$16 \div 2 = 8$; $8 \div 2 = 4$; $4 \div 2 = 2$, the starting value. This solution is checking.

2. *Read and understand the question.* This question is looking for the eighth
number in the pattern while the first five numbers are given.
Make a plan. Find the rule and continue the pattern to the eighth number.
Carry out the plan. Each of the numbers in the list is odd. The difference
between the first two numbers is 2, the difference between the second and
the third number is 4, the difference between the third and the fourth is 8,

and the difference between the fourth and the fifth number is 16. Each time, the difference between the terms doubles so that the sixth term would be 31 + 32 = 63; the seventh would be 63 + 64 = 127; and the eighth term would be 127 + 128 = 255. The eighth term would be 255.

Check your work. Another explanation of the rule is that the difference between two consecutive numbers in the list is always one more than the smaller number. For example, the number following 31 is equal to 31 + (31 +1) = 31 + 32 = 63. This is another way to extend this pattern to eight terms. Thus, the seventh term is 63 + (63 + 1) = 127, and the eighth term is 127 + (127 + 1) = 255.

3. *Read and understand the question.* This problem is looking for the number of handshakes for a group of people, and each person will shake hands with another exactly one time. It is understood that a person would not shake hands with himself or herself.

Make a plan. One approach to this question is to make a table of values. In the first column, list the number of people at the meeting and in the second column, list the number of handshakes that would take place. Start the table with one person, and look for a pattern from there.

Carry out the plan. If there is only one person at the meeting, there will not be any handshakes. If there are two people at the meeting, there will be one handshake. If there are three people, then the first person and second person shake hands, the second and third person shake hands, and the first and third person shake hands. This is a total of 3 handshakes. Examine these values in a table vertically.

Number of people	Number of handshakes
1	0
2	1
3	3
4	6

As you read down, the number of people increases by 1 each time. In the second column, the number of handshakes increases by adding one more than the previous increase. For example, add 1, then add 2, then add 3, and so on. To complete the table, use this pattern: 0 + 1 = 1, 1 + 2 = 3, so 3 + 3 = 6. There are 6 handshakes among 4 people.

Check your work. In order for each person to shake hands with each other person, person 1 would shake hands with person 2, person 3, and person 4. This is a total of 3 handshakes so far. Person 2 would now have to shake hands with person 3 and person 4, for an additional 2 handshakes. Person 3 would then need to shake hands with person 4 to complete the hand-shakes. This is a total of $3 + 2 + 1 = 6$ handshakes, which is the same con-clusion you drew by making the table.

4. *Read and understand the question.* You are looking for the missing value in a table where most of the numbers are given.

Make a plan. Look for a pattern in the table to help you figure out the miss-ing number. This pattern can be a horizontal (across) or vertical (up/down) pattern.

Carry out the plan. In this table, each of the values is filled in except for the location of the question mark, so look for a pattern with the other num-bers. For each given x-value, the corresponding y-value is one more than double the x-value. Check each row to make sure the pattern holds true for all numbers in the table. By using this pattern, the missing number is $2 \times 5 + 1 = 11$.

Check your answer. Another way to view this table is vertically. As the y-values increase, each number is two more than the previous one. By using this method, the missing number is $9 + 2 = 11$, which is the same answer you found by the other method.

Practice 2

1. *Read and understand the question.* The question is asking for the total num-ber of possibilities when selecting one flavor of ice cream and a cone or a dish. There are 3 different flavors and 2 ways to serve the ice cream.

Make a plan. Make a list of all of the combinations of one flavor with a cone or a dish. Then count the total number of possibilities to find the answer.

Carry out the plan. Make an organized list pairing each flavor with a cone, and then each flavor with a dish. An organized list could appear as follows:

cone—chocolate	dish—chocolate
cone—strawberry	dish—strawberry
cone—vanilla	dish—vanilla

This is a total of 6 different pairs; therefore, there are 6 different possibilities.

Check your work. Since there are three flavors paired with a cone or a dish, there should be $3 + 3 = 6$ possibilities. This is the same as the number of items in the organized list.

2. *Read and understand the question.* The question is asking for the total number of possibilities when selecting from 3 classes for period 1 and 4 classes for period 2.

Make a plan. Make a list of all of the combinations of the period 1 choices with each of the period 2 choices. Then, count the total number of possibilities to find the answer.

Carry out the plan. Make an organized list pairing the period 1 choices with each period 2 choice. An organized list could appear as follows:

History—Art	Math—Art	English—Art
History—Music	Math—Music	English—Music
History—Health	Math—Health	English—Health
History—Science	Math – Science	English—Science

This is a total of 12 different pairs; therefore, 12 different possibilities.

Check your work. Since there are 3 period 1 classes being paired with 4 different period 2 classes, there should be $4 + 4 + 4 = 12$ possibilities. This is the same as the number of items in the organized list.

3. *Read and understand the question.* The question is asking for the total number of possibilities when selecting the one type of roll and one topping. There are 2 different types of rolls and 4 toppings from which to choose.

Make a plan. Make a list of all of the combinations of each type of roll with one topping. Then, count the total number of possibilities to find the answer.

Carry out the plan. Make an organized list pairing each type of roll with each topping. An organized list could appear as follows:

plain—ketchup	sesame—ketchup
plain—mustard	sesame—mustard
plain—mayonnaise	sesame—mayonnaise
plain—relish	sesame—relish

This is a total of 8 different pairs; therefore, there are 8 different possibilities.

Check your work. Since there are 2 types of rolls being paired with 4 different toppings, there should be $4 + 4 = 8$ possibilities. This is the same as the number of items in the organized list.

the strategies of working backward and using logical reasoning

You can only find truth with logic if you have already found truth without it.

—G.K. CHESTERTON (1874–1936)

Mathematics and logic go hand in hand; they are related in many ways. Logical reasoning is used to solve many mathematical problems, and it is also used to explain many mathematical properties. You really cannot have one without the other. This idea will be highlighted in this lesson, which is based on the strategies of showing you how to work backward and use logical reasoning.

SOMETIMES, THE BEST APPROACH to solving a problem is to begin with the information at the end of a question and work back through the problem. This strategy of working backward will be explained, and you will have the opportunity to practice questions using this strategy. Using logical reasoning is another way to solve word problems, and two specific problem types using reasoning tables and Venn diagrams will be discussed in this section. The practice problems that follow will allow you to apply your skills using each of these strategies.

WORKING BACKWARD

When you are reading some types of math word problems, the information given in the question can seem like an endless list of facts. In questions such as this, it is sometimes helpful to begin with the last detail given. To apply this technique, start with the last piece of information, and work backward through the problem. Then, check your result by starting at the beginning of the question and working through the operations in the correct order.

Note that you will sometimes need to do the opposite, or inverse, operations when using this strategy. For example, if in the question, 2 was added to a number to get 5 as a result, you will need to subtract $5 - 2$ to get the beginning value of 3.

Read through the example that follows, which uses the strategy of working backward. Pay close attention to the inverse operations that are used when you are solving the problem.

Example

Ella sold cookies at a bake sale. During the first hour, she sold one-third of the number of cookies she brought. During the next hour, she sold 10 more. At the end of the sale, she sold half of the remaining cookies. If 10 cookies did not sell, how many cookies did she bring to the sale?

Read and understand the question. The question is asking for the total number of cookies that she began with at the start of the sale. Details about the number of cookies sold during each hour are given throughout the question.

Make a plan. Use the strategy of working backward. Begin with the fact that she had 10 cookies left over at the end of the sale, and work backward from there.

Carry out the plan. Start with the 10 cookies that were left over. At the end of the sale, she sold half of the remaining cookies. Thus, the other half is equal to the 10 cookies left over. Multiply 2 by 10 to double this amount and get 20. Before this, she sold 10 more, so add $20 + 10 = 30$. During the first hour, she sold one-third of the number of cookies. This means that she had two-thirds remaining, which is equal to 30. If two-thirds is 30, then one-third is 15. Add $30 + 15 = 45$. She began the sale with 45 cookies.

Check your work. Begin with the solution of 45 cookies and work forward through the question. During the first hour, she sold one-third of the cookies.

One-third of 45 is 15, and 45 – 15 = 30. During the next hour, she sold 10 more cookies: 30 – 10 = 20. At the end of the sale, she sold half of the remaining cookies. Half of 20 is equal to 10, so 20 – 10 = 10. She had 10 cookies remaining at the end of the sale. The solution is checking.

> **TIP:** Be sure to use the inverse (opposite) operations if necessary when using the strategy of working backward.

The questions in the practice set that follows can be simplified by using the strategy of working backward. Try this strategy for each, and then use the answer explanations at the end of the lesson to help refine this method of finding a solution.

PRACTICE 1

1. A child is playing with blocks. There are twice as many green blocks as red blocks. The number of blue blocks is five more than the number of red blocks. If there are 15 blue blocks, how many blocks of each color are there?

2. Sheila attended the local fair. She spent $5.50 playing games and spent double that amount on rides. She then spent $2.25 on an ice cream cone. If she had $4.25 left over, how much money did she bring to the fair?

3. On the first day of harvesting vegetables, a farmer picked three-fourths of his ears of corn. On the second day, he harvested half of the ears that were left. On the third day he harvested the remaining 50 ears. How many total ears of corn did he harvest?

USING LOGICAL REASONING

Two common strategies you can use in logical reasoning questions are Venn diagrams and reasoning with a table. Each strategy is demonstrated in the following examples.

Example 1: Using a Venn Diagram

Of 65 ninth graders at a high school, 40 take Russian and 30 take German. If 18 students take both Russian and German, how many ninth graders do not take either language?

Read and understand the question. The question is asking for the number of ninth graders who do not take Russian or German. The total number of ninth grade students in the school, as well as the number who take Russian, German, or both, is given.

Make a plan. Use a Venn diagram to answer this question. The Venn diagram should have a circle for the number of students who take German, which overlaps with a circle for the number of students who take Russian. This overlapping section represents the number of students who take both languages.

Carry out the plan. Start the diagram by drawing a rectangle that represents all of the ninth graders, and then place the two overlapping circles within the rectangle. The diagram could appear like the one shown here.

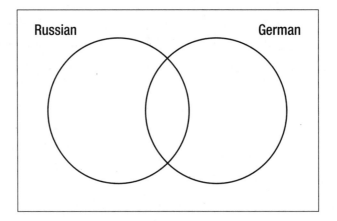

Because the number of students who take both is 18, place the 18 in the overlapping section between the circles. The 40 students taking Russian represent all of the students taking the language, including the 18 who take both languages. Therefore, the number in the part of the circle for Russian that does not overlap is $40 - 18 = 22$. In the same way, the circle for the 30 students taking German also includes the students taking both languages. So, the value in the part of the circle for German that does not overlap should be $30 - 18 = 12$. The Venn diagram should now look like the following figure.

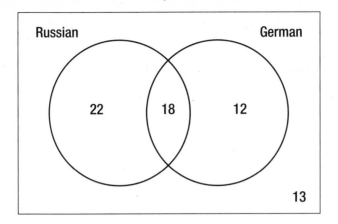

Now, add the three values in the diagram to get the total number of students who take one or both of the languages: 22 + 18 + 12 = 52. Subtract this amount from the total number of ninth graders at the school: 65 – 52 = 13. Thirteen students do not take either language. The number 13 is placed inside the rectangle but outside of either circle, as shown in the figure.

Check your work. Add the number for each category to make sure that the total number of ninth graders is 65. The number of students who take Russian only is 22, the number of students who take German only is 12, the number of students who take both languages is 18, and the number of students who do not take either language is 13: 22 + 12 + 18 + 13 = 65. Since each ninth grader at the school is in one of these four categories, this solution is checking.

....

TIP: When solving problems using a Venn diagram, be sure to subtract the number of objects that include more than one category from the total number in the category. If this is not done, the objects will be counted more than once in the problem.

....

Example 2: Reasoning with a Table

Tim, Curt, and Kara are brothers and sister in the same family. Kara is younger than Curt. Tim is not the youngest or the oldest. Assuming none of the children are twins or triplets, what is the order from youngest to oldest?

Read and understand the question. The question is asking for the birth order of three children from the same family. The children are not twins or triplets.

Make a plan. Make a table that includes the three people in the question along the side. Then, make one column for the youngest child, one for the middle child, and one column for the oldest child. Use the clues in the question to place an "X" in any box where the possibility can be eliminated. Use the process of elimination to figure out the order the children were born.

Carry out the plan. Start by making the table. A possible table follows.

	Youngest child	Middle child	Oldest child
Tim			
Curt			
Kara			

Use the clue that Kara is younger than Curt. With this information, you can eliminate the fact that Kara is the oldest, so place an "X" in this box. You can also reason that Curt is not the youngest, so place an "X" in this box. The table at this point could look like the one that follows.

	Youngest child	Middle child	Oldest child
Tim			
Curt	X		
Kara			X

The second clue states that Tim is not the youngest or the oldest, so place an "X" in both of these boxes. Tim is the middle child. Therefore, the youngest child must be Kara and Curt is the oldest child. The completed table could look like the following one.

	Youngest child	Middle child	Oldest child
Tim	X	✓	X
Curt	X	X	✓
Kara	✓	X	X

Check your work. Check your solution that Kara is the youngest, Tim is the middle child, and Curt is the oldest. This satisfies the clue that Kara is younger than Curt. It also is consistent with the fact that Tim is not the oldest or the youngest. This leaves Curt as the oldest child in the family. This solution is checking.

Using a Venn diagram or using logic with a table can be very helpful when you are solving these types of math word problems. Work through the questions in the practice set that follows. Apply these strategies to help find a solution, and be sure to check your answers.

PRACTICE 2

1. At a middle school, there are 75 students in drama club and 80 students in ski club. If 35 students participate in both drama club and ski club, how many students participate in only ski club?

2. A survey was given to Mrs. Miller's class about the pets they have. Of the 25 students in the class, 10 have a dog, 12 have a cat, and 5 do not have either pet. How many students in the class have both a dog and a cat?

3. Peter, Brenda, and Mike are three siblings in the same family and each likes a different color: red, blue, and green. Brenda's brother likes red. Blue was Mike's favorite color, but now he likes green. Which color does each person like?

ANSWERS

Practice 1

1. *Read and understand the question.* The question is asking for the number of blocks of each color when clues are given about the number of each.
   Make a plan. Use the strategy of working backward to solve this problem. Start with the fact that there are 15 blue blocks and use the inverse (opposite) operations when necessary.
   Carry out the plan. Because there are 15 blue blocks, and the number of blue blocks is 5 more than the number of red blocks, there are 15 − 5 = 10 red blocks. There are twice as many green blocks as red blocks. Therefore, there are 10 × 2 = 20 green blocks. There are 15 blue blocks, 10 red blocks, and 20 green blocks.

Check your work. Work forward through the question to check your work. Start with 20 green blocks. This is twice as many as the number of red blocks, so the number of red blocks is 10. The number of blue blocks is 5 more than the number of red blocks, so 10 + 5 = 15 blue blocks. Since this was given in the question, this solution is checking.

2. *Read and understand the question.* The question is asking for the total amount of money Ella had before she went to the fair.

Make a plan. Start with the amount of money left over, and use the strategy of working backward to find out how much she had at the start of the fair.

Carry out the plan. Begin with the $4.25 left over. Since she spent $2.25 on an ice cream cone, add $4.25 + $2.25 = $6.50. She also spent $5.50 on games, so add $6.50 + $5.50 = $12.00. She spent twice as much on rides as she did on games, so she spent $5.50 × 2 = $11.00 on rides. Now, add $12.00 + $11.00 = $23.00, which is the total amount of money she brought to the fair.

Check your work. Work forward through the question to check your work. She started with $23.00 and then spent $5.50 on games. $23.00 − $5.50 = $17.50. She spent twice as much on rides as on games, so subtract: $17.50 − $11.00 = $6.50. She then bought an ice cream cone for $2.25: $6.50 − $2.25 = $4.25. Since this is the amount of money she had left over, this answer is reasonable.

3. *Read and understand the question.* The question is looking for the total number of ears of corn harvested after he worked for three days.

Make a plan. Start with the fact that the farmer harvested 50 ears the third day, and use the strategy of working backward to the first day.

Carry out the plan. Start with 50 ears the third day. On the second day, he harvested half of the ears that were left, and 50 ears remained. This means that there were 100 ears on the second day and he harvested 50 of them. On the first day, he harvested three-fourths of the total number of ears. Since there were 100 ears on the second day, this represents one-fourth of the number of ears. If 100 is equal to one-fourth, then 300 is equal to three-fourths: 100 + 300 = 400 total ears were harvested.

Check your work. Start with the solution and work forward through the question to check your work. He began with 400 ears in the field. The first day he harvested three-fourths, so 300 of the 400 ears: 400 − 300 = 100 ears remain. The second day, he harvested half of what was left, so 50 ears: 100 − 50 ears remain. The final day he harvested the remaining 50 ears. This answer is checking.

Practice 2

1. *Read and understand the question.* The question is asking for the number of students that participate in only ski club. The total number of students in

drama club, the total number of students in ski club, and the number of students in both groups is given.

Devise a plan. Use a Venn diagram to answer this question. The Venn diagram should have a circle for the number of students in drama club that overlaps with a circle for the number of students in ski club. This overlapping section represents the number of students who are involved in both activities.

Carry out the plan. Start the diagram by drawing a rectangle that represents all of the students in the school, and then place the two overlapping circles within the rectangle. The diagram could appear like the one that follows.

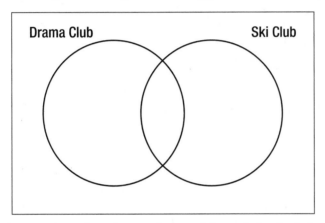

Because the number of students who participate in both clubs is 35, place the 35 in the overlapping section between the circles. The 75 students in drama club represent all of the students in drama club, including the 35 who are in both clubs. Therefore, the number in the part of the circle for drama club that does not overlap is 75 − 35 = 40. In the same way, the 80 students in ski club also include the students in both groups. So, the value in the part of the circle for ski club that does not overlap should be 80 − 35 = 45. The Venn diagram should now look like the figure that follows.

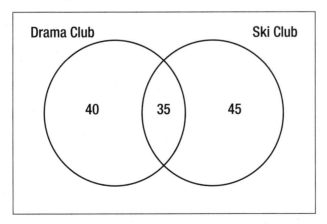

Thus, the value in the part of the circle for ski club that does not overlap is the number of students in the ski club only. There are 45 students in only the ski club, and not in both clubs.

Check your work. Add the two numbers that make up each circle to check for accuracy. There are 40 students in drama club only, and 35 students in both clubs. This is a total of $40 + 35 = 75$ students in the drama club. There are 45 students in ski club only and 35 students in both clubs for a total of $45 + 35 = 80$ students in the ski club. This solution is checking.

2. *Read and understand the question.* The question is asking for the number of students who have both a dog and a cat. The number of students in the class, along with the number of students without a cat or a dog, the number who have a dog, and the number who have a cat are given.

Make a plan. Use a Venn diagram to answer this question. The Venn diagram should have a circle for the number of students who have a dog that overlaps with a circle for the number of students who have a cat. This overlapping section represents the number of students who have both pets.

Carry out the plan. Start the diagram by drawing a rectangle that represents all of the students in the class, and then place the two overlapping circles within the rectangle. The diagram could appear like the one that follows.

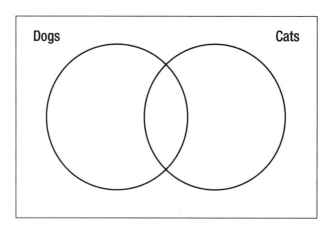

Because the number of students who have neither pet is 5, place a 5 in the diagram inside the rectangle but outside of both circles. Then, subtract $25 - 5 = 20$. This is the number of students who have one pet or both pets. Now, add the number of students who have either or both pets: $10 + 12 = 22$. Because this is two more than 20, two students were counted in both categories. This means that 2 students have both a cat and a dog. So, the value in the overlapping section should be 2. Thus, there are $10 - 2 = 8$ students who have a dog only and $12 - 2 = 10$ students who have a cat only. The completed Venn Diagram should now look like the figure that follows.

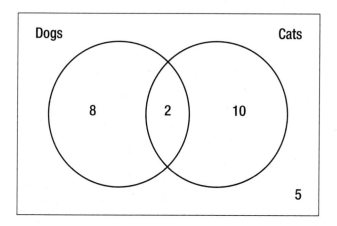

Check your work. Add the number for each category to make sure that the total number of students in the class is 25. The number of students who do not have either pet is 5, the number who have a dog only is 8, the number with a cat only is 10, and the number with both a cat and a dog is 2: $5 + 8 + 10 + 2 = 25$. Because there are 25 students in the class, this solution is checking.

3. *Read and understand the question.* The question is asking for the favorite color of three different people. Each person likes a different color of red, blue, or green.

Make a plan. Make a table that includes the three people in the question along the side. Then, make one column for red, one for blue, and one column for green. Use the clues in the question to place an "X" in any box where the possibility can be eliminated. Use the process of elimination to figure out the favorite color for each person.

Carry out the plan. Start by making the table. A possible table follows.

	Red	Blue	Green
Brenda			
Mike			
Peter			

Use the clue that Brenda's brother likes red. With this information, you can eliminate the fact that Brenda likes red, so place an "X" in that box.

The second clue states that Mike used to like blue, but now green is his favorite, so place an "X" in both blue and red for Mike. Therefore, Peter's favorite color must be red and Brenda's favorite is blue. The completed table could look like the one that follows.

	Red	Blue	Green
Brenda	✕	✓	✕
Mike	✕	✕	✓
Peter	✓	✕	✕

Check your work. Check your solution that Mike likes green, Brenda likes blue, and Peter likes red. This satisfies the clue that Mike used to like blue, but his favorite now is green. It also is consistent with the fact that Brenda's brother likes red, so Peter likes red. Then Brenda's favorite color is blue. This solution is checking.

the strategies of making a table and drawing a picture

Most people spend more time and energy going around problems than in trying to solve them.
—HENRY FORD (1863–1947)

At this point, you have seen that by using certain strategies, even difficult word problems can become easier to solve. The strategies presented in this lesson will focus on making a table and drawing a picture. Since "a picture is worth a thousand words," this seems like a great place to start.

DO NOT WAIT any longer. Use your time and energy wisely, and rely on the strategies presented in this lesson to solve the word problems that lie ahead.

MAKING A TABLE

Many times, the data in a problem needs to be organized in order for you to find a potential relationship. One way to efficiently arrange this information is in a table format. This way, if there are certain patterns in the numbers, you will be more likely to see them. This method also allows you, the problem solver, to better organize larger groups of information, as seen in the following example.

Example

At a carnival, it costs $2 to play each game and $3 for each ride ticket. Sandy spent $17 at the carnival. What is the largest number of ride tickets she could have purchased?

Read and understand the question. The question is seeking the largest number of ride tickets that could be bought when $17 was spent on both games and rides. Each game costs $2 and each ride costs $3.

Make a plan. Make a table to organize the information and to solve this problem. Have one row in the table for ride tickets and one row for game tickets. Then, use the table to find the total number of tickets bought when the total cost is $17.

Carry out the plan. Make the table. An example of the table is shown next.

	Number of Tickets	Amount Spent
Ride tickets		
Game tickets		
Total		

To complete the table, divide $17 by 3 to find the largest number of ride tickets that could have been purchased: $17 \div 3 =$ approximately 5.6666. Put a 5 in the column for the number of ride tickets. The total money spent on ride tickets: $5 \times \$3 = \$15$. Since Sandy now has $17 - \$15 = \$2$ left over, she must have spent this amount on games. Now, fill in the row for games using this leftover money. A possible completed table is shown next.

	Number of Tickets	Amount Spent
Ride tickets	5	$5 \times 3 = \$15$
Game tickets	1	$1 \times 2 = \$2$
Total	6	$17

Use the data in the table to check to see if the amounts add to $17: $15 + \$2 = \$17$.

Check your work. To check for the largest number of rides, try a larger number such as 6 for the number of rides. If Sandy bought 6 ride tickets, this would be $3 × 6 = $18, which is over the amount of money she spent.

....

TIP: When using the strategy of making a table, put the labels to the left in the table and make a row for each category. Then, you can use the table to work across and down in columns, depending on the question.

....

Example

The question that follows involves a series of numbers in a pattern. Read through the problem and answer explanation to see how using a table can make this pattern clearer and the solution easier to find.

Karl betters his time swimming in an event by 3 seconds in each of the 5 races of the season. If his time for the first race of the season is 58 seconds, what is his best time at the end of the season?

Read and understand the question. This question is looking for Karl's best time at the end of the season. Each time he betters his time, subtract 3 seconds from the previous time.

Make a plan. Start with his time at the start of the season, and then subtract 3 seconds for each of the 5 races. At the end of the 5 races, this is his time at the end of the season.

Carry out the plan. Make a table that includes his five races of the season, and his time for the first race of 58 seconds. Fill out the table, subtracting 3 seconds for every subsequent race. Continue this for all five races. The table could look like the following one.

Race Number	Time (Seconds)
1	58
2	55
3	52
4	49
5	46

The time for race 5, the final race of the season, is 46 seconds.

Check your work. Work backward to check this problem. Begin with his best time of 46 seconds and add 3 seconds until you reach his time for race 1: 46 + 3 + 3 + 3 + 3 = 58 seconds. This was his time at the start of the season, so this answer is reasonable.

Try the strategy of making a table with the following practice set. Check your solutions with the answer explanations at the end of the lesson to monitor your progress with this strategy.

PRACTICE 1

1. Harry has $0.14 in his pocket. How many different combinations of coins could he have in his pocket if there are at least two different types of coins (pennies, nickels, and dimes)?

2. Kevin is saving his money for a trip. Each day he saves $2 more than the previous day. If he begins by saving $9 on the first day, how much money has he saved after 5 days?

DRAWING A PICTURE

The strategy of drawing a picture to solve a word problem may also shed light on patterns that may occur. It also gives a bird's-eye view to what is going on in certain problems. Take, for example, the following question. The method of solution will give a picture, or diagram, about the locations and distances used in the question that will make finding the correct answer much easier.

Example

Joe leaves school and travels 3 miles directly west to his home. Later that night, he leaves home and goes north 2 miles to the public library. When he has finished at the library, he travels 5 miles east to his friend's house and then 2 miles south to his grandmother's house. If his grandmother lives 2 miles from his school, how far away is his grandmother's house from his house?

Read and understand the question. This question asks for the distance between Joe's house and his grandmother's house.

Make a plan. Joe's route after school is detailed in the problem. Draw a picture of the route Joe took after school. Use these details, along with knowledge of east, west, north, and south to find the distance between points.

Carry out the plan. Draw a picture of Joe's route. Recall that on a map, west is to the left of north, north is to the left of east, east is to the right of north, and south is to the right of west. Use these facts as you retrace the path. A possible picture of Joe's path is shown in the following figure.

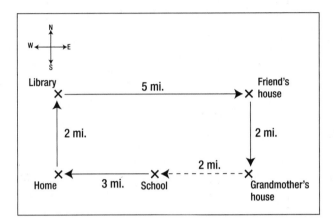

His path is 3 miles west, then 2 miles north, 5 miles east, and 2 miles south. At this point, he is 2 miles directly east of school. Because he lives 3 miles east of school, his home is 2 + 3 = 5 miles from his grandmother's house. The distance between Joe's house and his grandmother's house is 5 miles.

Check your work. Compare the distances to check this problem. Joe went 2 miles north and 2 miles south, so these two distances cancel each other out. He went 3 miles west, and 5 miles east. This is a difference of 2 miles. However, since his trek did not begin at his house, add 3 miles between his home and school: 2 + 3 = 5 miles. This solution is reasonable.

....

TIP: When you are drawing a picture or diagram to solve a problem, try to label in the figure each of the details given in the question. That way, important information will not be left out when you are trying to find a solution.

....

Drawing a picture or diagram can be a very useful tool when you are solving certain math word problems. Use this strategy when you are working out the problems in the practice set that follows.

PRACTICE 2

1. Tyrone is planting seeds in his garden. Each seed needs a total of one square foot to have enough room to grow. How many seeds can be planted if the dimensions of his garden are 1 yard by 2 yards?

2. Tony, Betty, Joe, and Frank are sitting in a row of four chairs facing toward you. Betty is not sitting on the end. Joe is next to Tony, but is not sitting next to Betty. Frank is sitting one seat to the right of Betty. In what order are the four people in the row?

3. Robert has a rectangular yard. He is going to place a post every 4 feet to put up a fence for his dog, which will include one at each corner. If the dimensions of his yard are 24 feet by 12 feet, how many posts will he need?

ANSWERS

Practice 1

1. *Read and understand the question.* This question is asking for the number of combinations of coins Harry could have in his pocket if he has $0.14. *Make a plan.* Make a table of possibilities using pennies, nickels, and dimes. Add the amounts to be sure each has a total of $0.14. There are at least two different types of coins in Harry's pocket. *Carry out the plan.* Make a table with a row for dimes, nickels, and pennies. A possible table is shown next.

	First combination	Second combination	Third combination
Number of pennies			
Number of nickels			
Number of dimes			
Total amount			

Fill in the table considering all possible combinations of two or more types of coins. For example, start with 1 dime. With 1 dime, he could have

4 pennies. Continue to fill in the table in this fashion. A possible completed table follows.

	First combination	Second combination	Third combination
Number of pennies	4 = $0.04	9 = $0.09	4 = $0.04
Number of nickels	0	1 = $0.05	2 = $0.10
Number of dimes	1 = $0.10	0	0
Total amount	$0.14	$0.14	$0.14

There are three possible combinations of coins.

Check your work. The only ways to have $0.14 in change with at least two different coins are 1 dime and 4 pennies, 2 nickels and 4 pennies, and 1 nickel and 9 pennies. There are three combinations, so this problem is checking.

2. *Read and understand the question.* Kevin is saving money beginning the first day with $9. This question asks for the total amount saved after 5 days.
Make a plan. Make a table with a column for the number of days from 1 to 5, and a column for the total amount saved each day. Add the totals in the amount saved column to find the total money saved.
Carry out the plan. Make the table. A possible table follows.

Day	Amount of money saved ($)
1	
2	
3	
4	
5	

Fill in the amounts for each day by beginning with $9 on day 1. Add $2 for each day, and continue to day 5. A completed table follows.

Day	Amount of money saved
1	$9
2	$11
3	$13
4	$15
5	$17

Add the amounts in the total amount saved column to find the total money: $9 + 11 + 13 + 15 + 17 = 65$. He has saved $65 for his trip.

Check your work. Start with the total amount saved, $65, and subtract the amounts saved each day. $\$65 - 9 - 11 - 13 - 15 - 17 = 0$. This question is checking.

Practice 2

1. *Read and understand the question.* You are asked to find the total number of seeds that can be planted in a garden. The dimensions of the garden are given in yards and the square units given for the seeds are in square feet. *Make a plan.* Draw a picture of the rectangular garden, but be sure to use the smaller unit of square feet: 1 yard = 3 feet. Convert yards to feet by multiplying by 3 for each dimension.

   Carry out the plan. Draw a picture of the garden. Because the dimensions are 1 yard by 2 yards, label the diagram 3 feet by 6 feet. Each seed requires 1 square foot to grow, so divide the garden into square feet as shown in the following figure.

   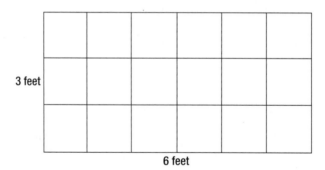

   Since this garden can be divided into 18 square feet, 18 seeds can be planted.

   Check your work. Check your work by finding the area of the rectangle in square feet. *Area = length × width*, so $A = 3$ feet $\times 6$ feet $= 18$ square feet.

2. *Read and understand the question.* This question asks for the placement of four people in a row of chairs, given clues about where they are sitting. *Make a plan.* Use the strategy of drawing a picture to find where each is sitting. For example, use four lines in a row to represent the four chairs and label each as the information is given.

   Carry out the plan. Start by drawing the lines to represent each chair. A possible picture follows.

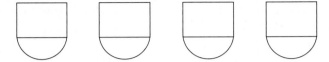

Use the first clue that Betty is not on the end. Therefore, she is sitting in one of the two middle chairs. The next clue states that Joe is next to Tony, but is not next to Betty. Thus, Joe must be sitting on the end not next to Betty, and Tony is between them. The diagram could now appear as follows.

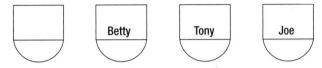

The final clue states that Frank is sitting one chair to the right of Betty, which is the only chair left. The order of the people is Frank, Betty, Tony, Joe.
Check your work. Read through the clues to be sure the final order checks. Betty is not on the end in this order, and is not next to Joe. Frank is to the right of Betty. This solution is checking.

3. *Read and understand the question.* This question is asking for the number of fence posts needed for a rectangular yard. The posts are placed every 4 feet.
Make a plan. Draw a picture of the yard and make a fence post every 4 feet. Count the posts in the diagram to find the total.
Carry out the plan. Draw a picture of the yard. A possible picture could look like the following one.

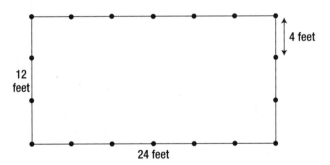

The sides labeled 24 feet are divided into 6 sections, and the sides labeled 12 feet are divided into 3 sections. Counting each of the posts, there are 18 posts needed.
Check your work. When you are dividing 24 feet into 4-foot sections, there are actually 7 posts needed on these sides for a total of $7 + 7 = 14$ posts. This includes the posts on each corner. Two additional posts are needed on each 12-foot side to complete the yard: $14 + 2 + 2 = 18$ total posts. The solution is checking.

the strategies of solving a simpler problem and guess and check

*Leaders are problem solvers by talent
and temperament, and by choice.*
—HARLAN CLEVELAND (1918–2008)

You have tackled some challenging problems so far, each one made easier by using various strategies. In this lesson, the strategies of solving a simpler problem and guess and check will be two more to add to your list. Each strategy will be explained and examples given to help guide you. Remember, there are many ways to solve most problems. Having a variety of strategies to choose from can only help you on this road to word-problem solving success!

SOLVING A SIMPLER PROBLEM

With some word problems, the numbers used within the problem may be difficult to use. They may be too large for one of the previously mentioned strategies. Or, the numbers given in the question may be complicated, and should be simplified to make the question easier to handle. One way to manage this situation is to make the problem easier to solve by using simpler values than the ones given. Take, for example, the question that follows.

Example

In a certain school district, there are 294 students who are bused to school each day. If each bus can fit 52 students, how many buses are needed?

Read and understand the question. This question is looking for the number of buses needed for a certain number of students.

Make a plan. Divide the total number of students by the number of students that can fit on a bus. Because the numbers are not easily divided, make the numbers simpler by rounding. Then, a reasonable solution can be found much more easily.

Carry out the plan. Instead of using 294 students, use a value of 300. Since each bus can fit 52 students, round this value to 50. Divide 300 by 50 to get 6 buses.

Check your answer. Check the solution by multiplying 6 buses by 50, which is 300. Since there are only 294 students who need to ride the bus and each bus can actually fit 52 students, there will be extra seats left. The solution is checking.

....

TIP: Be careful when using rounding to make simpler, more compatible numbers. As in the question, the number of students was rounded up slightly and the number of students who would fit on each bus was rounded down. This way, the number of buses needed was not underestimated. Always check a solution to make sure it is reasonable.

....

Here is another example of making a problem simpler. In the question that follows, you are asked to find the sum of many numbers. Notice how finding the sum of a few numbers can lead to an easier way to find a solution.

Example

What is the sum of the first 19 natural numbers?

Read and understand the question. This question is asking for the total when the first 19 whole numbers beginning with 1 are added together.

Make a plan. This problem seems difficult when you are trying to add all 19 values. Begin by making the problem simpler by adding just a few numbers, and then look for a pattern.

Carry out the plan. You need to add the natural numbers from 1 to 19. This would appear as $1 + 2 + 3 + 4 + \ldots + 18 + 19$. However, instead of adding the numbers in order, begin by adding the first and last numbers: $1 + 19 = 20$. Continue this pattern using the next numbers: $2 + 18 = 20$, $3 + 17 = 20$, $4 + 16 = 20$, and so on until you reach $9 + 11 = 20$. The value 10 in the list will not have a paired value. Therefore, there are nine sums of 20, plus the number 10: $9 \times 20 + 10 = 180 + 10 = 190$. The sum is 190.

Check your work. One way to check your work is to add the numbers in the list by hand or with a calculator. Each of these methods results in a sum of 190. The solution is checking.

Try the problems in the following practice set to apply the strategy of solving a simpler problem.

PRACTICE 1

1. Charlotte has $62 to spend on new clothes. She would like to buy a shirt for $14.99, a pair of pants for $19.99, and a new pair of shoes for $24.99. If there is no sales tax on the items, does she have enough money to buy all three?

2. What is the sum of the first 10 even natural numbers?

GUESS AND CHECK

Guess and check, also known as trial and error or guess and test, is another important strategy. This is the method that many people choose when they cannot come up with any other approach to solve a problem. Although guess and check can be used on just about any question, the problems in this lesson lend themselves to this strategy.

The method of guess and check is exactly what it says: You guess an answer and then check to see if it works. It is important to note, however, if you get a correct answer on the first guess that you should show at least three trials each time this strategy is used. You should always show at least three guesses, two incorrect and, of course, the correct answer.

Read through the following sample to see an example of how to use the strategy of guess and check.

Example

Gail has 13 coins for a total of $1.00. If she only has dimes and nickels, how many of each coin does she have?

Read and understand the question. This question is looking for the number of dimes and nickels that Gail has. She has a total of 13 coins that add to exactly $1.00.

Make a plan. Use the strategy of guess and check to solve this problem. Begin with a guess as to the number of nickels and number of dimes, and be sure that the number of coins adds to 13. Then, check to see if the total amount of money adds to $1.00.

Carry out the plan. Start your guesses with 6 dimes and 7 nickels. Six dimes is equal to $0.60 and 7 nickels is equal to $0.35. This is a total of $0.60 + $0.35 = $0.95. This guess is too low.

For the next guess, try a greater amount of dimes. Try 8 dimes and 5 nickels. This is a total of $0.80 + $0.25 = $1.05. This guess is too high.

For the next guess, try 7 dimes and 6 nickels. Seven dimes is equal to $0.70 and 6 nickels is equal to $0.30. This is a total of $0.70 + $0.30 = $1.00. This is the correct answer.

Check your work. Be sure that the solution meets all of the facts in the question. Gail had only nickels and dimes for a total of 13 coins. Seven dimes ($0.70) and six nickels ($0.30) is a total of 13 coins, which add to $1.00. This answer is checking.

....

TIP: It is important to show at least three trials when you are using the strategy of guess and check. Show one guess greater than the right answer, one guess less than the right answer, and the correct answer. Always show all work to prove the solution works, and remember, it may take more than three trials to find the correct answer.

....

The following question set provides practice on using the strategy of guess and check. Try these problems and make use of this approach to make the problems easier to solve.

PRACTICE 2

1. A person spots a number of birds and squirrels in a park. Each squirrel has 4 feet, each bird has two feet, and there are a total of 24 feet among all of the animals. If there are 3 more squirrels than birds, how many of each animal is in the park?

2. Hannah is half as old as her aunt. The sum of their ages is 42. How old is Hannah?

3. Henry ate two less than triple the number of candies that Noah ate. If they ate a total of 30 candies, how many did each person eat?

ANSWERS

Practice 1

1. *Read and understand the question.* This question asks if there is enough money to buy three items. The total amount of money Charlotte has to spend and the price of each item is given.
   Make a plan. Make the problem simpler by adjusting the values. Use rounding to make the numbers easier to use.
   Carry out the plan. Round the price of the shirt from $14.99 to $15.00, round the price of the pants from $19.99 to $20.00, and round the price of the shoes from $24.99 to $25.00. Add the rounded amounts of each item: $15 + $20 + $25 = $60. She has enough money to pay for the items.
   Check your work. When adding the rounded amounts, she has $62 − $60 = $2 left over. Because each of the prices was rounded up, she would actually have slightly more than $2.00 left over.

2. *Read and understand the question.* This question is looking for the sum of the first 10 even numbers. The even numbers are 2, 4, 6, 8 . . .
   Make a plan. This problem seems difficult when you are trying to add all 10 values. Begin by making the problem simpler by adding just a few numbers, and then look for a pattern.
   Carry out the plan. You need to add the first 10 even numbers. This would appear as 2 + 4 + 6 + 8 + 10 + 12 + 14 + 16 + 18 + 20. However, instead of adding the numbers in order, begin by adding the first and the second-to-last number: 2 + 18 = 20. Continue this pattern using the next numbers: 4 + 16 = 20, 6 + 14 = 20, 8 + 12 = 20. The values of 10 and 20 in the list will

not have a paired value. Therefore, there are four sums of 20, plus the number 10 and the number 20: $4 \times 20 + 10 + 20 = 80 + 10 + 20 = 110$. The sum is 110.

Check your work. One way to check your work is to add the numbers in the list by hand or with a calculator. Each of these methods results in a sum of 110.

Practice 2

1. *Read and understand the question.* In this question, you are seeking the number of birds and the number of squirrels spotted in a park. The fact that there are 3 more squirrels than birds is given. It is also given that there are 24 feet between the animals.

   Make a plan. Use the strategy of guess and check to solve this problem. Show at least three guesses for the number of birds and number of squirrels, and be sure that the number of squirrels is 3 more than the number of birds.

   Carry out the plan. Begin with the guess of 4 squirrels and 1 bird. This would be a total of $4 \times 4 = 16$ feet for the squirrels and $2 \times 1 = 2$ feet for the bird. This is a total of 18 feet. This guess is too low.

   For the next guess, try 6 squirrels and 3 birds. This would be a total of $6 \times 4 = 24$ feet for the squirrels and $3 \times 2 = 6$ feet for the birds. This is a total of 30 feet, which is too high.

   Now, try a guess of 5 squirrels and 2 birds. This would be a total of $5 \times 4 = 20$ feet for the squirrels and $2 \times 2 = 4$ feet for the birds. This is a total of 24 feet, which is the correct answer.

   Check your work. Check your answer to be sure that it fits all of the information in the question. There are a total of 24 feet among the animals, and there are 3 more squirrels than birds. Five squirrels is 3 more than 2 birds, and there are a total of $20 + 4 = 24$ feet among them. This answer is checking.

2. *Read and understand the question.* This question is looking for Hannah's age. She is half as old as her aunt. If you add their ages, the sum is 42.

   Make a plan. Apply the strategy of guess and check to solve this problem. Try a number for Hannah's age and then double it to represent her aunt's age. See if the sum of the two ages is equal to 42.

   Carry out the plan. The sum of their ages is 42, and half of 42 is 21. Since you are going to be adding a number and double that same number, begin your guesses with a value somewhat smaller than 21. Try a guess of 12 for Hannah's age. Her aunt's age would then be $12 \times 2 = 24$. The sum of these ages is $12 + 24 = 36$. This is too low.

Try the next guess of 16 for Hannah's age. Her aunt's age would then be $16 \times 2 = 32$. The sum of these ages is $16 + 32 = 48$. This is too high.

For the next guess, use 14 for Hannah's age. Her aunt's age would then be $14 \times 2 = 28$. The sum of these ages is $14 + 28 = 42$. This is the correct answer.

Check your work. Check to make sure that the answer fits all of the information given in the question. Since Hannah is half as old as her aunt, 14 is half of 28. The sum of their ages is 42, so check the sum of their ages: $14 + 28 = 42$. This answer is checking.

3. *Read and understand the question.* You are looking for the number of candies each person ate. The total number of candies is given, along with the fact that Henry ate two less than triple the number that Noah ate.

Make a plan. Use the strategy of guess and check. Start with an amount for Noah, then triple this amount and subtract 2. Make sure that the amounts for each person add to 30.

Carry out the plan. There are a total of 30 candies, so half of this amount is 15. Start the first guess below this amount.

Try 10 candies for Noah. Then, Henry ate $10 \times 3 - 2 = 30 - 2 = 28$ candies. This is a total of $10 + 28 = 38$ candies. This guess is too high.

Try 5 candies for Noah. Then, Henry ate $5 \times 3 - 2 = 15 - 2 = 13$ candies. This is a total of $5 + 13 = 18$ candies. This guess is too low.

Try 7 candies for Noah. Then, Henry ate $7 \times 3 - 2 = 21 - 2 = 19$ candies. This is a total of $7 + 19 = 26$ candies. This guess is too low, but very close to the correct answer. Remember, do at least three trials for each problem, but it may take more to find the solution.

Try 8 candies for Noah. Then, Henry ate $8 \times 3 - 2 = 24 - 2 = 22$ candies. This is a total of $8 + 22 = 30$ candies. This is the correct answer.

Check your work. Check your work by making sure that your answer fits the information given in the question. Henry ate two less than triple the amount that Noah ate. Since Noah ate 8 candies, two less than triple this amount is $24 - 2 = 22$ candies. The total number of candies was 30, and $22 + 8 = 30$. This answer is checking.

ⓢ ⓔ ⓒ ⓣ ⓘ ⓞ ⓝ 2

number sense word problems—making them make sense

COUNTING IS ONE of the first mathematical processes we did as small children. Looking back, learning how to count was the key to beginning to learn about math and its operations. Keeping this in mind, remember that math word problems are really just regular math problems, but your task is to figure out which operation or set of operations to use in order to solve the problem. Other areas to tackle are the many different types of numbers involved in word problems, such as fractions, decimals, and percents, to name a few. These questions can be made easier by knowing the basics of how to work with each number type. This section will outline these number types and the ins and outs of working with them. By applying the strategies outlined in the previous lessons, and the helpful tips and refreshers in this section, you can conquer number sense word problems!

This section details word problems including specific number sets and the tools and strategies to use when solving these types of problems.

This section will introduce you to math word problems including:

- basic numbers and operations
- integers
- fractions
- decimals
- percents

- applications of percent
- ratios
- rate
- proportion
- scale

LESSON

basic number and integer word problems

Mathematics is the supreme judge;
from its decisions there is no appeal.
—Tobias Dantzig (1884–1956)

This section will focus on counting and the other basic operations in math, and will apply them to solving math word problems.

LET'S BEGIN OUR study of basic number and integer word problems with a review of the number sets and the rules for the four basic math operations.

Rational Numbers: any number that can be expressed as $\frac{a}{b}$, where b is not equal to zero, and a and b are both integers.

This is the set that includes any number that can be written as a fraction, such as $5 = \frac{5}{1}$ and $\frac{5}{6}$, and it also includes all repeating and terminating decimals, like 0.5 and 1.3.

Integers: { . . . –2, –1, 0, 1, 2 . . . }

This is the set of whole numbers and their opposites. There are no fractions or decimals in this set.

Whole Numbers: {0, 1, 2, 3, 4 . . . }

This set contains no fractions or decimals, and no negative values.

Natural (Counting) Numbers: {1, 2, 3, 4, 5 . . . }

This is the set of whole numbers without zero.

The questions in this section will focus on the natural numbers, whole numbers, and integers. Lesson 7 will concentrate on the set of rational numbers in fraction form, and Lesson 8 will cover word problems with decimals.

INTEGERS

Integers are the set of whole numbers and their opposites, and there are no fractions or decimals contained in this set. There are rules that can be applied when performing operations with integers, and we use the concept of absolute value when applying these rules of arithmetic.

The **absolute value** of a number is the distance the number is away from zero on a number line. It is a positive value no matter if you are counting from the right or from the left.

Absolute value is announced by using vertical bars on either side of the number. To find the absolute value of a number, count the number of units it is away from zero on a number line.

For example, the absolute value of 4, written as $|4|$, is equal to 4. The absolute value of –5, written as $|-5|$, is equal to 5. Each of these examples is shown on the number line that follows.

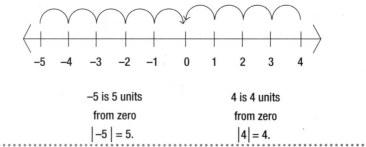

−5 is 5 units
from zero
$|-5| = 5.$

4 is 4 units
from zero
$|4| = 4.$

TIP: Think of the absolute value of a number as the value of the number without any sign. The absolute value of a number is always positive.

Adding Integers

1. If the signs are the same, add the absolute values and keep the sign.

    Examples: $2 + 3 = 5$ $-2 + -4 = -6$ $-10 + -20 = -30$

2. If the signs are different, subtract the absolute value of one number from the absolute value of the other and take the sign of the number with the larger absolute value.

    Examples: $-2 + 4 = 2$ $5 + -11 = -6$ $-21 + 30 = 9$

Subtracting Integers

Change the subtraction to addition and change the sign of the number following the subtraction to its opposite. Then, follow the rules for addition.

Examples: $-4 - 5$ $9 - (-6)$ $-14 - (-11)$

                                               $-4 + -5$ $9 + 6$ $-14 + 11$

                                               -9 15 -3

TIP: Subtracting is the same as adding the opposite. For example, $5 - 4 = 5 + -4$. Each has a value of 1.

Multiplying and Dividing Integers

1. If there is an even number of negative signs, perform the operation as usual and make the answer positive.

    Examples: $-8 \times -6 = 48$ $-20 \div -4 = 5$ $-2 \times -3 \times -4 \times -1 = 24$

2. If there is an odd number of negative signs, perform the operation as usual and make the answer negative.

    Examples: $-2 \times 5 = -10$ $40 \div -8 = -5$ $-10 \times -2 \times -4 = -80$

PRACTICE 1

Match each expression with the correct value.

$-1-5$	-4
$-7+-9$	-20
3×-4	7
$-14 \div -2$	-16
$6-10$	-6
$-10 \times -1 \times -2$	-12

Integer Word Problem Practice

Now that the rules for each of the operations with these number sets have been reviewed, it is time to move on to word problems that involve these operations. The sample questions that follow will use the steps to solving word problems introduced in the previous section to break down each question into smaller pieces.

Example 1

Jason is eating at a restaurant and orders a hamburger and a side of sliced apples. There are 12 slices of apple in the portion. If there are 5 calories in each apple slice and 230 calories in the hamburger, what is the total number of calories of his meal?

Read and understand the question. This question asks for the total number of calories in a meal. The number of calories in the hamburger and the number of calories in each piece of apple is given. The number of slices, 12, is also given.

Make a plan. The fact that you are finding the *total* number of calories in this case tells you to use addition. Find the total number of calories by adding the number of calories in the hamburger with the total number of calories in the apple slices. Find the total number of calories in the apple slices first by multiplying the number of calories in one slice by 12.

Carry out the plan. Find the total calories in the apple slices by multiplying 12 by 5.

$$12 \times 5 = 60$$

There are 60 calories in the side of apple slices. Now, add this amount with the number of calories in the hamburger: $60 + 230 = 290$. There are a total of 290 calories in the meal.

Check your answer. Check the solution to this problem by working backward. Start with the solution of 290 calories and subtract the number of calories in the hamburger:

$$290 - 230 = 60$$

This amount should be the total number of calories in the side of apple slices. Divide this number by 5 calories in each slice: $60 \div 5 = 12$ slices, which was given in the question. This solution is checking.

....

TIP: The key word *total* does not always signal addition. Use the context of the problem to figure out which is the correct operation.

....

Example 2

On four plays, a football team has a gain of 8 yards, a loss of 10 yards, a gain of 2 yards, and then a loss of 20 yards. If they began on the 50-yard line, what yard line were they on at the end of the four plays?

Read and understand the question. This question deals with the gain and loss of yards of a football team. Each gain is a positive number and should be added; each loss is a negative number and should be subtracted. The starting place of the team is given, and you are looking for the yard line at the end of the plays.

Make a plan. Begin with the number 50. This represents the 50-yard line, or halfway across the football field. Use the rules of adding and subtracting integers to find the yard line on which the team ends the plays.

Carry out the plan. Start with the number 50 to represent the 50-yard line and add a gain of 8 yards: $50 + 8 = 58$. Then, subtract a loss of 10 yards: $58 - 10 = 48$. This represents the 48-yard line of the original team. Now, add a gain of 2 yards: $48 + 2 = 50$. Then, subtract a loss of 20 yards: $50 - 20 = 30$. The team is on their own 30-yard line at the end of the four plays.

Check your answer. Check your answer by working backward and use inverse operations. Begin with the solution of the 30-yard line. Add 20 yards, subtract 2 yards, add 10 yards, and subtract 8 yards: $30 + 20 - 2 + 10 - 8 = 50$. Since this is where the team started before the plays, this answer is checking.

Example 3

Gretchen is working on the problem –30 divided by –6 at her desk at school. What should she write for the correct answer?

Read and understand the question. This question is looking for the result after a division problem. The two numbers in the problem are both negative.

Make a plan. Use the rules for dividing integers to complete this question.

Carry out the plan. Divide $-30 \div -6 = 5$. There is an even number of negatives in the problem, so the answer is positive.

Check your answer. Check your answer by doing the inverse operations. Multiply 5 by –6 and the result is –30. This question is checking.

PRACTICE 2

1. Sue and Terry each have marbles. Sue has five more than twice the number that Terry has. If Terry has 8 marbles, how many does Sue have?

2. Zach scores 27 points in a basketball game. If he made one 3-point basket and every other basket he made scored 2 points, what is the total number of baskets he made?

3. The low temperature on Monday is –9 degrees. If the low temperature on Tuesday is 5 degrees below this value, what is the low temperature on Tuesday?

4. A diver descended 20 feet into the water from a boat at the surface of a lake. She then rose 12 feet, and then descended another 18 feet. At this point, what is her depth in the water?

5. Jackson divided –350 and –7 correctly. What was the quotient?

6. What is the product of –4, –2, and –5?

ANSWERS

Practice 1

$$-1 - 5 = -6$$
$$-7 + -9 = -16$$
$$3 \times -4 = -12$$
$$-14 \div -2 = 7$$
$$6 - 10 = -4$$
$$-10 \times -1 \times -2 = -20$$

Practice 2

1. *Read and understand the question.* This question is looking for the total number of marbles Sue has. The number that Terry has is given in the problem.
   Make a plan. Start with the fact that Terry has 8 marbles and use key words
   and basic operations to solve. Calculate what number is equal to *five more
   than twice eight marbles.*
   Carry out the plan. Five more than twice eight marbles is equal to 8 times 2,
   plus 5.
   $$8 \times 2 + 5 = 16 + 5 = 21$$
   Sue has 21 marbles.
   Check your answer. Since Sue has 21 marbles, this is five more than twice
   the number of marbles Terry has. Work backward: subtract 5 and divide
   by 2:
   $$21 - 5 = 16$$
   $$16 \div 2 = 8$$
   This is the number Terry has, so this solution is checking.
2. *Read and understand the question.* This question is looking for the total
   number of baskets scored in a basketball game. Each basket Zach made
   scores either three or two points.
   Make a plan. In this problem, subtract the points made by the 3-point basket. Then, divide the total number of points left by two.

Carry out the plan. Zach scored 27 points. Subtract $27 - 3 = 24$ to take out the one 3-point basket. Then, divide 24 by 2 to get twelve 2-point baskets. He scored a total of 13 baskets.

Check your answer. To check this problem, add one 3-point basket to twelve 2-point baskets. $12 \times 2 + 3 = 24 + 3 = 27$ total points. This solution is checking.

3. *Read and understand the question.* This question gives the low temperature on Monday, and asks you to find the low temperature on Tuesday.

Make a plan. The temperature on Tuesday is 5 degrees below the low temperature on Monday. Subtract 5 from –9 to find the answer.

Carry out the plan. Subtract. Be sure to use the rules for subtracting integers by adding the opposite: $-9 - 5 = -9 + -5 = -14$. When you are adding, if the signs are the same, add the absolute values and keep the sign. The low temperature on Tuesday is –14 degrees.

Check your answer. To check this problem, do the opposite operation. Be sure to follow the rules for adding integers. Add $-14 + 5$. The signs are different, so subtract and take the sign of the number with the larger absolute value: $-14 + 5 = -9$. This was the temperature on Monday, so this solution is checking.

4. *Read and understand the question.* This question asks for the depth of a diver in the water after she dived in at sea level.

Make a plan. Start with the value of zero for sea level, then add or subtract as the diver descends (goes down) or rises.

Carry out the plan. Start at 0. She descends 20 feet, so $0 - 20 = -20$ feet. She rose 12 feet, so add 12 to –20: $-20 + 12 = -8$. She then descends another 18 feet, so subtract $-8 - 18 = -26$ feet. Her depth at this point is –26 feet, or 26 feet below sea level.

Check your work: $0 - 20 + 12 - 18 = -20 + 12 - 18 = -8 - 18 = -26$. This problem is checking.

5. *Read and understand the question.* This question is looking for the result of a division problem. The key words are *divided* and *quotient.*

Make a plan. The quotient is the result of a division problem. Divide the two numbers in the question and be aware of the signs of the numbers.

Carry out the plan. Divide –350 by –7 by dividing their absolute values. The result is 50. Because there are two negatives being divided, the final result is positive 50.

Check your answer. Check your answer by multiplying: 50×-7 is –350. There is only one negative sign in this check so the product is negative. This solution is checking.

6. *Read and understand the question.* This question is looking for the product of three values.

Make a plan. The key word is *product,* which means multiply. Be aware of the number of negative numbers being multiplied.

Carry out the plan. Multiply the absolute values of the three numbers together. There are an odd number of negative signs (3), so the result is negative: $-4 \times -2 \times -5 = -40$.

Check your answer. Check your answer by multiplying $4 \times 2 \times 5 = 40$. There are an odd number of negatives, so the result is -40. This solution is checking.

fraction word problems

Arithmetic is numbers you squeeze from your head to your
hand to your pencil to your paper till you get the answer.
—CARL SANDBURG (1878–1967)

This lesson will provide practice performing operations with fractions, as well as
strategies that can be used when you are solving word problems with fractions.

IF YOU LOOK UP the definition of the set of rational numbers, you may get
a description like the one mentioned in the previous lesson. They are the set of
numbers that can be expressed as $\frac{a}{b}$, where b is not equal to zero, and a and b are
both integers. This is just a complicated description of very familiar numbers
known as fractions. Before embarking on our study of fractions and word problems involving fractions, let's review a few key concepts.

FACTORS, MULTIPLES, GCFs, AND LCMs

Factor: A number that divides into another number without
leaving a remainder. For example, the factors of 12 are 1, 2, 3, 4, 6,
and 12.

Greatest Common Factor (GCF): The largest value that divides
each of the terms without a remainder.

Example: Find the greatest common factor of 18 and 24.

The factors of 18 are 1, 2, 3, 6, 9, and 18.

The factors of 24 are 1, 2, 3, 4, 6, 8, 12, and 24.

The factors 1, 2, 3, and 6 are common to both lists, but the greatest common factor is 6.

Multiple: The result of multiplying a number by another whole number. For example, multiples of 4 are $4 \times 1 = 4$, $4 \times 2 = 8$, $4 \times 3 = 12$, and so on.

Least Common Multiple (LCM): The smallest value that each term divides into without leaving a remainder.

Example: Find the least common multiple of 12 and 30.

Multiples of 12 are 12, 24, 36, 48, 60, 72 . . .

Multiples of 30 are 30, 60, 90 . . .

The smallest common number in each list is 60, so 60 is the least common multiple of 12 and 30.

. . . .

TIP: To help remember the difference between factors and multiples, think factors *fit* into a number, and you *multiply* to get multiples.

. . . .

SIMPLIFYING FRACTIONS

We learn to simplify fractions in order to make them easier to use. If the numbers are smaller, they are usually less difficult to use.

To simplify a fraction, divide the numerator (top) and the denominator (bottom) by the greatest common factor. For example, take the fraction $\frac{16}{20}$. The greatest common factor of the numerator and the denominator is 4. Divide each value by 4 to simplify the fraction to its lowest terms: $\frac{16 \div 4}{20 \div 4} = \frac{4}{5}$.

An **improper fraction** has a numerator whose absolute value is greater than or equal to the absolute value of its denominator: $\frac{4}{3}$ is an improper fraction.

A **mixed number** is a number made up of a whole number part and a fraction part. The number $5\frac{2}{7}$ is a mixed number.

CHANGING IMPROPER FRACTIONS TO MIXED NUMBERS

To change an improper fraction into a mixed number, divide the numerator by the denominator. Then, place the remainder, if there is one, over the same denominator. For example, to change $\frac{12}{7}$ to a mixed number, divide 12 by 7. The result is 1, with a remainder of 5. The mixed number is $1\frac{5}{7}$.

CHANGING MIXED NUMBERS TO IMPROPER FRACTIONS

To change a mixed number into an improper fraction, multiply the whole number part by the denominator. Then, add this product to the numerator. This value is the new numerator, and the denominator remains the same. For example, to change the mixed number $4\frac{1}{3}$ to an improper fraction, multiply 4 by 3 to get 12. Then, add $12 + 1 = 13$ to get the new numerator. Because the denominator remains the same, the improper fraction is $\frac{13}{3}$.

PRACTICE 1

Fill in each answer blank with the correct solution.

1. The greatest common factor of 8 and 32 is _____.

2. The least common multiple of 15 and 25 is _____.

3. The fraction $\frac{24}{36}$ in simplest form is equal to _____.

4. The improper fraction $\frac{11}{9}$ is equivalent to the mixed number _____.

5. The mixed number $3\frac{2}{5}$ written as an improper fraction is equivalent to _____.

ADDING AND SUBTRACTING FRACTIONS

To add or subtract fractions, you must first have a common denominator. Then, add or subtract the numerators and keep the denominators. For example, in order to add $\frac{1}{7} + \frac{3}{7}$, add the numerators and keep the denominator of 7. The result is $\frac{4}{7}$.

On the other hand, to subtract $\frac{3}{4} - \frac{1}{2}$, use the least common multiple of the denominators to find the least common denominator. The least common multiple of 4 and 2 is 4, so change the second fraction to also have a denominator of 4. To do this, multiply the numerator and denominator by 2: $\frac{1}{2} = \frac{1 \times 2}{2 \times 2} = \frac{2}{4}$. Now that there is a common denominator, subtract the numerators and keep the denominator. $\frac{3}{4} - \frac{2}{4} = \frac{1}{4}$.

When you are adding or subtracting mixed numbers, add or subtract the fraction parts and add or subtract the whole number parts. For example, to add $2\frac{1}{5} + 1\frac{2}{5}$, the fraction part adds to $\frac{1}{5} + \frac{2}{5} = \frac{3}{5}$ and the whole number parts add to $2 + 1 = 3$. The answer is $3\frac{3}{5}$.

....

TIP: Use the greatest common factor (GCF) to simplify fractions, and use the least common multiple (LCM) to find a common denominator.

....

MULTIPLYING FRACTIONS

To multiply fractions, multiply across the numerators and denominators. Then, simplify the product if necessary. For example, $\frac{3}{5} \times \frac{4}{9} = \frac{12}{45}$. Since 12 and 45 share a greatest common factor of 3, divide each by 3 to simplify the fraction. The simplified fraction is $\frac{4}{15}$.

....

TIP: If the numerator of a fraction has a common factor with the denominator of a fraction in a multiplication problem, you may cross cancel the common factors. For example, when multiplying $\frac{3}{5} \times \frac{4}{9}$, 3 and 9 share a common factor of 3. By canceling out this factor, the problem becomes $\frac{\overset{1}{3}}{5} \times \frac{4}{\underset{3}{9}} = \frac{4}{15}$. The fraction is in simplest form.

....

To multiply mixed numbers, first change to improper fraction form, and then follow the same steps as before. For example, $2\frac{1}{2} \times \frac{3}{5} = \frac{\overset{1}{5}}{2} \times \frac{3}{\underset{1}{5}} = \frac{3}{2} = 1\frac{1}{2}$.

DIVIDING FRACTIONS

Dividing fractions is very similar to multiplying fractions. When you are dividing, change the problem so that you are multiplying by the reciprocal of the divisor. Then, multiply the fractions as usual. For example, in order to divide $\frac{4}{21} \div \frac{2}{7}$, first change the problem to multiplication by multiplying by the reciprocal. The reciprocal of a fraction switches the numerator and the denominator. For example, the reciprocal of $\frac{2}{7}$ is $\frac{7}{2}$. The problem becomes $\frac{4}{21} \times \frac{7}{2}$. Now cross cancel the common factors and multiply across: $\frac{\overset{2}{\cancel{4}}}{\underset{3}{\cancel{21}}} \times \frac{\overset{1}{\cancel{7}}}{\underset{1}{\cancel{2}}} = \frac{2}{3}$. As in multiplication, when you are dividing mixed numbers, change them to improper fraction form first and then follow the same steps as before.

PRACTICE 2

1. Subtract: $\frac{6}{13} - \frac{4}{13}$

2. Add: $2\frac{1}{3} + 3\frac{1}{6}$

3. Multiply: $\frac{10}{21} \times 1\frac{3}{4}$

4. Divide: $\frac{3}{8} \div \frac{7}{9}$

FRACTION WORD PROBLEMS

Now it is time to apply this knowledge of fractions and operations with fractions to math word problems. Use the following sample question as an example, and then try the practice questions that follow to test your skills.

Example

Jamie needs $\frac{3}{4}$ yards of material for her school project. If she will bring in enough material for herself and three classmates, how much material does she need altogether?

Read and understand the question. Jamie needs material for herself and three others, so she needs four times the number of yards for one person.

Make a plan. Use key words to help solve this problem. The key word altogether in this context suggests multiplication. Multiply the number of yards for one person by 4 to find the total amount.

Carry out the plan. $\frac{3}{4} \times 4 = \frac{3}{4} \times \frac{4^1}{1} = \frac{3}{1} = 3$. Jamie needs a total of 3 yards of material for herself and three classmates.

Check your answer. One way to check the result is to divide the total amount needed by the number of people. Three yards divided by 4 people $= 3 \div 4 = \frac{3}{4}$ yard per person. This solution is checking.

PRACTICE 3

1. Phil is making two different types of cookies. For one recipe, he needs $\frac{2}{3}$ cups of sugar and for the other recipe he needs $\frac{3}{4}$ cups of sugar. What is the total amount of sugar he needs for both recipes?

2. Ally measured a board to be $12\frac{1}{2}$ feet long. If she cuts off a piece that measures $5\frac{1}{4}$ feet long, what is the length of the remaining piece?

3. Becky is building a square pen for her pet rabbit. The directions for the pen call for a $10\frac{2}{5}$ foot-long chain link fence that will surround the pen. How long is each side of the pen?

4. Thirty students in Mr. Joyce's room are working on projects. The first day he gave them $\frac{3}{5}$ of an hour to work. On the second day, he gave them half as much time as the first day. How much time did the students have altogether to work on their projects?

ANSWERS

Practice 1

1. 8
2. 75
3. $\frac{2}{3}$
4. $1\frac{2}{9}$
5. $\frac{17}{5}$

Practice 2

1. Subtract the numerators and keep the denominator: $\frac{6}{13} - \frac{4}{13} = \frac{2}{13}$.
2. Change to a common denominator of 6: $2\frac{1}{3} + 3\frac{1}{6} = 2\frac{2}{6} + 3\frac{1}{6} = 5\frac{3}{6} = 5\frac{1}{2}$.
3. Change to improper fraction form first, then cross cancel common factors.
   $\frac{10}{21} \times 1\frac{3}{4} = \frac{10}{21} \times \frac{7}{4} = \frac{5}{6}$
4. Divide by multiplying by the reciprocal: $\frac{3}{8} \div \frac{7}{9} = \frac{3}{8} \times \frac{9}{7} = \frac{27}{56}$.

Practice 3

1. *Read and understand the question.* This question is looking for the total amount of sugar needed for two cookie recipes.
   Make a plan. The key word *total* in this context tells you to add. Add the amounts for each recipe to find the solution.
   Carry out the plan. Add $\frac{2}{3} + \frac{3}{4}$ by first finding a common denominator. The least common multiple of 3 and 4 is 12, so the problem then becomes $\frac{8}{12} + \frac{9}{12} = \frac{17}{12}$. Change this amount to a mixed number. The final answer is $1\frac{5}{12}$ cups.
   Check your answer. Check your answer by subtracting one of the values from the total: $1\frac{5}{12} - \frac{2}{3} = \frac{17}{12} - \frac{8}{12} = \frac{9}{12} = \frac{3}{4}$, which was the other value being added. This answer is checking.
2. *Read and understand the question.* The question is asking for the length of a board after it has been cut. The length of the board before the cut and the amount cut off are given in the problem.
   Make a plan. Because you are looking for the length of the remaining piece, or the amount left over, subtract the amount cut off from the original length.
   Carry out the plan. Use subtraction and your knowledge of fractions to carry out this plan. Subtract $12\frac{1}{2} - 5\frac{1}{4}$ by getting a common denominator of 4 and subtracting the whole number parts and the fraction parts: $12\frac{1}{2} - 5\frac{1}{4} = 12\frac{2}{4} - 5\frac{1}{4} = 7\frac{1}{4}$. The remaining board is $7\frac{1}{4}$ feet long.
   Check your answer. Check your answer by adding. $7\frac{1}{4} + 5\frac{1}{4} = 12\frac{2}{4} = 12\frac{1}{2}$. This problem is checking.
3. *Read and understand the question.* This problem is asking for the length of a square pen when the total length of all four sides is given.
   Make a plan. Divide the total length of the chain link fence by the four equal sides of the square pen.

Carry out the plan. Divide $10\frac{2}{5}$ by 4 equal sides. Be sure to change to improper fraction form and multiply by the reciprocal:

$$10\frac{2}{5} \div 4 = \frac{52}{5} \div \frac{4}{1} = \frac{\overset{13}{\cancel{52}}}{5} \times \frac{1}{\cancel{4}_1} = \frac{13}{5} = 2\frac{3}{5}.$$

Each side of the pen is $2\frac{3}{5}$ feet long.

Check your answer. Check your answer by working backward. Multiply the length of one side by 4 to find the total length. $2\frac{3}{5} \times 4 = \frac{13}{5} \times \frac{4}{1} = \frac{52}{5} = 10\frac{2}{5}$. This problem is checking.

4. *Read and understand the question.* This problem asks for the total amount of time the students had to work on their projects in class, given time on two different days.

Make a plan. Be careful of the extra information in the problem. The fact that there are 30 students in the class is not necessary to solve this problem. The key word *altogether* tells you to find the amount for each day and add the times together.

Carry out the plan. Find the amount for the first day. $\frac{3}{5}$ of an hour is equal to $\frac{3}{5} \times 60$ minutes $= \frac{3}{{}_1\cancel{5}} \times \frac{\overset{12}{\cancel{60}}}{1} = 36$ minutes. The second day is half this amount: $\frac{1}{2}$ of $36 = \frac{1}{{}_1\cancel{2}} \times \frac{\overset{18}{\cancel{36}}}{1} = 18$ minutes. To find the total, add $36 + 18 = 54$ minutes.

Check your answer. Check this answer by making sure the answer fits the details in the question. They worked three-fifths of an hour the first day, which was 36 minutes, then half that amount, or 18 minutes the second day. This was a total of 54 minutes, so this answer is checking.

decimal word problems

How many times can you subtract 7 from 83, and what is left afterwards? You can subtract it as many times as you want, and it leaves 76 every time.

—AUTHOR UNKNOWN

Certain situations can seem complicated, but underneath, they are really not that difficult. This lesson will focus on the basics of decimals and decimal word problems, so that they will not seem as difficult the next time around.

DECIMAL PLACE VALUE

Decimals are based on place value. Each place value is named by the distance it is from the decimal point. Commonly used decimal places are shown in the following figure.

> **TIP:** Remember, decimals and fractions are very closely related. Take
> these examples:
>
> The decimal 0.3, which is *three tenths*, is equal to $\frac{3}{10}$.
> The decimal 0.21, which is *twenty-one hundredths*, is equal to $\frac{21}{100}$.
> The decimal 0.127, which is *one hundred twenty-seven thousandths*,
> is equal to $\frac{127}{1,000}$.

ROUNDING DECIMALS

To round a decimal to a certain place value, look to the decimal place to the right
of the place that you are rounding. If the digit to the right is 4 or less, keep the
number in the place value and drop the numbers to the right. For example, to
round the number 2.354 to the nearest hundredth, look to the number 4 in the
thousandths place. Because this number is 4 or less, keep the 5 in the hundredths
place and drop the digit to the right. The rounded decimal becomes 2.35.

To round the number 6.178 to the nearest tenth, look to the 7 in the hundredths place. Because this number is 5 or greater, round the 1 in the tenths place
to 2 and drop the digits to the right. The rounded number becomes 6.2.

> **TIP:** Remember that our money system is based on dollars and cents.
> When rounding with money, always round to the nearest hundredth,
> unless told otherwise.

ADDING AND SUBTRACTING DECIMALS

To add or subtract decimals, line up the decimal points and then add or subtract
as you would whole numbers. For example, to add 12.26 + 14.11, line up the decimal points vertically and add.

$$
\begin{array}{r}
12.26 \\
+\ 14.11 \\
\hline
26.37
\end{array}
$$

TIP: When you are adding or subtracting, if one number goes out more decimal places than another, trailing zeros can be added to the right of the decimal point to make the problem easier. For example, when adding 2.3 and 4.15 you may line up the numbers as follows and add a zero after the 3 in 2.3:

```
  2.30
+ 4.15
  6.45
```

As always, be sure to line up the decimal points when adding and subtracting.

MULTIPLYING DECIMALS

To multiply decimals, multiply the decimals as you would whole numbers. Then, count the number of decimal places in each number being multiplied to find the total number of decimal places. Move the decimal this many places from the right in the answer to find the correct decimal place.

Multiply 6.23 by 5.4.

```
     6.23
   × 5.4
     2492
 + 31150
    33.642
```

Because there are two decimal places in the first value and one in the second, move the decimal point 2 + 1 = 3 places from the right for the final answer. The final answer is 33.642.

DIVIDING DECIMALS

To divide decimals, move the decimal point in the divisor to the right so that the divisor is a whole number. Then, move the decimal point in the dividend the same number of places. Line up the decimal points vertically so that the quotient has a decimal point lined up with the dividend.

To divide 13.44 by 2.1, set up the division and change 2.1 to the whole number 21 by moving the decimal point one place to the right. Then, move the decimal point of the dividend one place to the right to make 13.44 into 134.4. Next, divide 134.4 by 21 as shown in the following figure.

$$
\begin{array}{r}
6.4 \\
2.1\overline{)13.44} \\
\underline{-12\ 6\downarrow} \\
84 \\
\underline{-84} \\
0
\end{array}
$$

The quotient is 6.4.

PRACTICE 1

1. Round 18.649 to the nearest hundredth.

2. Subtract: 648.99 − 21.48.

3. Add: 45.6 + 32.05.

4. Multiply: 2.8×6.51.

5. Divide: $230.3 \div 23.5$.

DECIMAL WORD PROBLEMS

The following examples use the basic properties of decimals in math word problems. Read and work through each problem to get an idea of how to approach decimal word problems, and then try your hand at the practice problems following them. Do not worry; answer explanations are given to help you through any rough spots.

Example 1

Ted and Kristen are shopping for pens and notebooks. The pens cost $0.50 each and the notebooks are $1.50 each. Ted buys 2 pens and 1 notebook and pays with a $5 bill. Kristen buys 3 pens and 2 notebooks and pays with a $10 bill. What is the total amount of change they get back?

Read and understand the question. This problem asks for the total amount of change that two people get back while shopping for pens and notebooks. The cost of each pen and notebook is given, along with the money they have to spend.

Make a plan. Use the strategy of making a table to keep track of how much each person spent. Then, subtract this amount from the money they used to pay for the items to find how much change they got back. Add the amounts of change to find the total.

Carry out the plan. Make a table of the amounts spent.

Person	Amount Spent on Pens	Amount Spent on Notebooks	Total
Ted	2 pens = $1.00	1 notebook = $1.50	$2.50
Kristen	3 pens = $1.50	2 notebooks = $3.00	$4.50
		Total amount of money spent is	$7.00

Because they paid with a $5.00 bill and a $10.00 bill for a total of $15.00, they get back $15 − $7 = $8 in change.

Check your answer. To check your answer, figure out the total number of pens and notebooks bought, and subtract this amount from $15. A total of 5 pens were bought for a total cost of $2.50. A total of 3 notebooks were purchased for a total of $4.50. This is a total of $2.50 + $4.50 = $7.00, leaving $15 − $7 = $8 in change between both people. This solution is checking.

Example 2

The individual times for four runners on the same team in a track relay event were 29.3 seconds, 35.8 seconds, 41.1 seconds, and 31.2 seconds. What was the total time for the team?

Read and understand the question. This question is looking for the total time it took a team to finish a race. The individual times are given.

Make a plan. Add each of the individuals' times together to find the total amount of time for the relay team.

Carry out the plan. Add the times vertically and line up the decimal points.

$$
\begin{array}{r}
29.3 \\
35.8 \\
41.1 \\
+\ 31.2 \\
\hline
137.4
\end{array}
$$

The team completed the race in 137.4 seconds.

Check your answer. Check your answer by subtracting each number from 137.4 until you reach zero seconds: $137.4 - 29.3 - 35.8 - 41.1 - 31.2 = 0$. This problem is checking.

PRACTICE 2

1. Andy had a balance of $65.58 in his bank account. He then paid a bill for $42.14. How much money did he have left in his account after he paid the bill?

2. Elise bought 12 cans of dog food. The price for the cans was 3 for $4.69. How much money did she spend on dog food?

3. At a movie theater, a box of popcorn cost $4.50, and a soda cost $2.00. If Sandra bought exactly 5 items for her family and spent $15.00, what did she buy?

4. Tim swam four laps of the pool in 140.8 seconds. If he swam at the same pace the entire time, how long did it take him to swim one lap?

5. Ken bought 12.5 gallons of gasoline at $3.12 a gallon. What is the total amount of money he spent on gasoline?

ANSWERS

Practice 1

1. 18.65
2. 627.51
3. 77.65
4. 18.228
5. 9.8

Practice 2

1. *Read and understand the question.* This problem asks for the total amount in a bank account after a bill has been paid. The amount in the account to start and the amount of the bill are given.
   Make a plan. The key phrase *how much does he have left* tells you to subtract the amount of the bill from the amount in the account.
   Carry out the plan. Subtract $65.58 − $42.14. Be sure to line up the decimal points. The answer is $23.44.
   Check your answer. Check your answer by working backward. Begin with the new balance of the account and add the amount of the bill to see if it is equal to the original amount in the bank account. $23.44 + $42.14 = $65.58. This answer is checking.

2. *Read and understand the question.* This question asks for the total amount of money spent on 12 cans of dog food. The price for 3 cans is given.
   Make a plan. Because $3 \times 4 = 12$, multiply the price for 3 cans by 4 to find the total amount spent on 12 cans.
   Carry out the plan: $4.69 \times 4 = $18.76. She spent a total of $18.76 on 12 cans.
   Check your answer. Check this solution by working backward. Divide $18.76 by 4. This is equal to $4.69, which is the cost of 3 cans. Four groups of 3 cans are equal to 12 cans, so this answer is checking.

3. *Read and understand the question.* This question is looking for the exact items that were bought when the price of each is given, along with the fact that there are exactly 5 items purchased for a total of $15.00.
   Make a plan. Use the strategy of guess and check and your knowledge of decimals to find the numbers of each item that were bought.
   Carry out the plan. Start with 1 box of popcorn and 4 sodas. Because the cost of soda is $2.00, then 4 sodas would be $8.00: $4.50 + $8.00 = $12.50, which is too low. You may now realize that an odd number of boxes of

popcorn cannot result in a whole dollar amount, so try even amounts of popcorn.

Try 4 boxes of popcorn and 1 soda. This is a total of 4 × $4.50 = $18, which is too high even before the 1 soda is added.

Try 2 boxes of popcorn and 3 sodas. This is a total of 2 × $4.50 = $9.00 and 3 × $2.00 = $6.00: $9.00 + 6.00 = $15.00. This is the correct solution.

Check your answer. Check your answer by making sure that the answer fits the details in the problem. She bought a total of 5 items for a total of $15.00, so this answer is checking.

4. *Read and understand the question.* This question is looking for the length of time it took to swim one lap when the time to swim 4 laps is given.

Make a plan. Divide the total time for the 4 laps by 4 to find out the time for 1 lap.

Carry out the plan. The total time was 140.8 seconds: 140.8 divided by 4 equals 35.2 seconds. At the same pace, it took Tim 35.2 seconds to complete 1 lap.

Check your answer. Check this solution by multiplying 35.2 by 4 laps: 35.2 × 4 = 140.8. This question is checking.

5. *Read and understand the question.* This question is looking for the total amount spent on gasoline. The number of gallons and the price per gallon are given.

Make a plan. Multiply the number of gallons by the price per gallon to find the total money spent on gasoline.

Carry out the plan. $3.12 multiplied by 12.5 is equal to $39.00. Ken spent a total of $39.00 on gasoline.

Check your answer. Check this result by using inverse operations. Divide the total amount spent by the price per gallon: $39.00 ÷ $3.12 = 12.5 gallons. This question is checking.

ratio and rate word problems

There is no branch of mathematics, however abstract, which may not someday be applied to the phenomena of the real world.
—NICOLAI LOBACHEVSKY (1792–1856)

Whether we enjoy them or not, ratios and rates can be found everywhere in our daily lives. Ratios help us to compare quantities of different things. When traveling between locations, we travel at a rate of speed. When we are shopping at a grocery store, the unit price tells us if the cost is reasonable. The information in this lesson will review information about ratios and rate and will give practice solving word problems on these topics.

RATIO

A ratio is a comparison of two numbers by division. There are three different ways to write a ratio: as a fraction, using a colon, and using the word *to*.

For example, if the ratio of cats to dogs is 1 to 3, this can also be written as 1:3 or $\frac{1}{3}$.

The order of the terms in the ratio is very important. For instance, in the example above, the word "cats" was mentioned first and the word "dogs" was mentioned second. This means that the first number in the ratio stands for cats and the second number stands for dogs. Knowing this concept is essential as you move forward into more ratios, rates, and eventually, proportions.

SIMPLIFYING RATIOS

Simplifying a ratio is basically the same process as simplifying a fraction. To do this, find the greatest common factor between the two numbers in the ratio and divide each by this common factor. For example, if the fans at a football game number 300 for the home team and 200 for the visiting team, this can be written as the ratio 300:200. The greatest common factor of each of these numbers is 100, so the ratio can be simplified to 3:2, 3 to 2, or $\frac{3}{2}$. This means there are 3 home team fans for every 2 visiting team fans.

TIP: Although they appear very similar to fractions, ratios are different from fractions. Fractions compare a part to a whole. In a ratio, the values could be comparing a part of a group to another part of a group, or part of a group to the entire group. For example, if there are 14 girls and 15 boys in a class, the ratio of girls to boys is $\frac{14}{15}$ but the ratio of girls to total students is $\frac{14}{29}$. Be aware of the labels of the ratio to help with this concept.

CONTINUED RATIOS

A continued ratio is a ratio that compares more than two numbers. For example, the ratio 2:3:4 is a continued ratio. In order to simplify a continued ratio, divide each term by the greatest common factor of all the terms.

To simplify the continued ratio 15:30:45, divide by the greatest common factor, 15. The ratio simplifies to 1:2:3.

PRACTICE 1

1. Write this ratio in two other ways: 4 to 5.

2. Write this ratio in two other ways: $\frac{1}{2}$.

3. Simplify the ratio: 33 to 44.

4. Simplify the ratio: $\frac{14}{21}$.

5. Simplify the continued ratio 20 to 40 to 50.

RATE

A **rate** is similar to a ratio, but it is specifically a comparison of two units, not only two numbers. Common rates used in everyday life are miles per hour, feet per second, gallons per minute, and basically any time when you are comparing two units.

. . . .

TIP: A **unit rate** is a rate where the second number is always one. For instance, traveling 100 miles in 2 hours is a rate, but simplifying this to traveling 50 miles in 1 hour is a unit rate.

. . . .

To find a unit rate, simply divide the quantity of the first unit by the quantity of the second unit. For example, if someone can type 200 words on a keyboard in 4 minutes, the unit rate is $\frac{200 \text{ words}}{4 \text{ minutes}}$ because $200 \div 4 = 50$.

PRACTICE 2

Find the unit rate for each.

1. $\dfrac{25 \text{ gallons}}{5 \text{ minutes}}$

2. $\dfrac{60 \text{ feet}}{10 \text{ seconds}}$

3. $\dfrac{600 \text{ miles}}{30 \text{ gallons}}$

RATIO AND RATE WORD PROBLEMS

Now that the concepts of ratio and rate have been reviewed, try the following word problems.

PRACTICE 3

1. There are 65 boys and 52 girls in the drama club. What is the ratio of boys to girls, in simplest form?

2. On a farm that has only two types of animals, there are 34 cows and 12 horses. What is the ratio of horses to the total number of animals on the farm, in simplest form?

3. Over a period of 3 hours, 180 leaves fell from a tree. At this rate, how many leaves fell in one hour?

4. Georgia drove a total of 252 miles and used 12 gallons of gasoline. What is this rate in miles per gallon?

5. Tyler scored 21 goals in 7 soccer games. At this rate, about how many goals did he score each game?

6. While climbing down a mountain, Anthony descended 45 feet every hour. At this rate, how many feet will he descend in 6 hours?

ANSWERS

Practice 1

1. $4:5; \frac{4}{5}$
2. 1 to 2; 1:2
3. 3 to 4. Divide by the greatest common factor, 11.
4. $\frac{2}{3}$. Divide by the greatest common factor, 7.
5. 2 to 4 to 5. Divide by the greatest common factor, 10.

Practice 2

1. $\dfrac{5 \text{ gallons}}{1 \text{ minute}}$
2. $\dfrac{6 \text{ feet}}{1 \text{ second}}$
3. $\dfrac{20 \text{ miles}}{1 \text{ gallon}}$

Practice 3

1. *Read and understand the question.* This question asks for the ratio of boys to girls in simplest form, so the answer will have to be in lowest terms.
Make a plan. Find the number of boys first and the number of girls second. Find the greatest common factor of the two values and divide by this number to simplify.
Carry out the plan. The ratio becomes $\frac{65 \text{ boys}}{52 \text{ girls}}$. The greatest common factor of these numbers is 13, so divide each by 13:
$$\frac{65 \text{ boys}}{52 \text{ girls}} = \frac{65 \div 13}{52 \div 13} = \frac{5 \text{ boys}}{4 \text{ girls}}$$
This can also be written as 5 boys:4 girls or 5 boys to 4 girls.
Check your answer. To check your result, work backward and multiply each value by 13. The values become 65 and 52, respectively, which were the original numbers in the problem. This answer is checking.

2. *Read and understand the question.* This question is looking for the ratio of the number of horses to the total number of animals on a farm. The number of horses and the number of cows is given in the question.
Make a plan. Compare the number of horses with the total number of animals by adding the number of cows and the number of horses together. Then, be sure that the answer is in simplest form by dividing by the greatest common factor of both numbers.
Carry out the plan. The number of horses is 12, and the total number of animals is $34 + 12 = 46$. The ratio can be written as $\frac{12 \text{ horses}}{46 \text{ animals}}$. The greatest common factor of both numbers is 2, so the simplified ratio is $\frac{6 \text{ horses}}{23 \text{ animals}}$. This can also be written as 6 horses:23 animals or 6 horses to 23 animals.
Check your answer. Be sure that you are comparing the correct quantities. This question asks for the number of horses compared to the total number of animals, which is the number of cows and horses added together. In addition, check to make sure that the value for horses is first, because it was mentioned first in the question. This solution is checking.

3. *Read and understand the question.* This question is looking for the unit rate, or how many leaves fell in 1 hour, when the number of leaves that fell in 3 hours is given.
Make a plan. Divide the number of leaves that fell in 3 hours by 3 to find the unit rate.
Carry out the plan: $\frac{180 \text{ leaves}}{3 \text{ hours}} = \frac{60 \text{ leaves}}{1 \text{ hour}}$
Check your answer. Multiply to check your answer. Sixty leaves per hour for 3 hours is equal to $60 \times 3 = 180$ leaves. This solution is checking.

4. *Read and understand the question.* This question is looking for the unit rate, or how many miles per gallon Georgia's car got, when the total miles driven and gallons used are given.

   Make a plan. Divide the number of miles by the number of gallons used to find the unit rate.

   Carry out the plan: $\dfrac{252 \text{ miles}}{12 \text{ gallons}} = \dfrac{21 \text{ miles}}{1 \text{ gallon}}$

   Check your answer. Multiply to check your answer. Twenty-one miles per gallon multiplied by 12 gallons is equal to $21 \times 12 = 252$ miles. This answer is checking.

5. *Read and understand the question.* This question is looking for the unit rate, or how many goals were scored in each game, when the total goals and number of games are given in the question.

   Make a plan. Divide the number of goals by the total number of games to find the unit rate per game.

   Carry out the plan: $\dfrac{21 \text{ goals}}{7 \text{ games}} = \dfrac{3 \text{ goals}}{1 \text{ game}}$

   Check your answer. Multiply to check your answer. Three goals per game for 7 games is equal to $3 \times 7 = 21$ goals. This solution is checking.

6. *Read and understand the question.* This question is looking for the total number of feet Anthony will descend when the unit rate is given.

   Make a plan. Multiply the unit rate by the total number of hours he is descending.

   Carry out the plan: 45 feet per hour multiplied by 6 hours is equal to $45 \times 6 = 270$ feet.

   Check your answer. Divide to check your answer.

   $\dfrac{270 \text{ feet}}{6 \text{ hours}} = \dfrac{45 \text{ feet}}{1 \text{ hour}}$

   This solution is checking.

proportion word problems

There is a difference between not knowing and not knowing yet.
—SHEILA TOBIAS (1935–)

One important ingredient in the recipe for success in solving math word problems
is persistence; another is practice. This lesson will concentrate on the concepts
of proportion and scale. Practice solving word problems on these topics is pro-
vided. Use this lesson to become more skilled in this important area of math problem solving by applying persistence and practice.

PROPORTIONS

While a ratio is a comparison of two numbers, and a rate is a comparison of two
units, a **proportion** is a comparison of two ratios.

....

TIP: Like ratios, proportions can be written in different ways. The most
common ways are using a colon *a:b = c:d* or fraction form $\frac{a}{b} = \frac{c}{d}$.

....

Although there are a number of ways to test if a proportion is a true proportion,
one way is to cross multiply. To cross multiply, find the product of the numerator of one ratio and the denominator of the other ratio. Then, set this product equal
to the result of multiplying the other numerator by the other denominator.

For example, use cross multiplication to see if the following proportion is a true proportion.

$$\frac{2}{3} = \frac{4}{6}$$

In this proportion, multiply 2×6 and 3×4. Each product is 12, so the proportion is a true proportion.

To test the proportion that follows, cross multiply 5×11 and 8×7.

$$\frac{5}{8} = \frac{7}{11}$$
$$5 \times 11 = 8 \times 7$$
$$55 \neq 56$$

The cross products are not equal, so this is not a true proportion.

....

TIP: Another name for cross multiplying is the **means-extremes property.** In the proportion $a{:}b = c{:}d$, the terms a and d are known as the extremes and b and c are known as the means.

....

SOLVING PROPORTIONS

In other cases, one or more terms of the proportion are unknown. These proportions can be solved by cross multiplying, or by the means-extremes property, in order to find the value or values to make it a true proportion. To do this, find the product of the means and set it equal to the product of the extremes. Then, solve the equation.

For example, to solve the proportion $\frac{x}{10} = \frac{6}{15}$, multiply the means, and set it equal to the product of the extremes. This is equal to $15x = 60$. Divide each side of the equation by 15:

$$\frac{15x}{15} = \frac{60}{15}$$
$$x = 4$$

The value $x = 4$ makes this a true proportion.

PRACTICE 1

Circle true or false for each statement that follows.

1. The proportion $\frac{5}{6} = \frac{15}{18}$ is a true proportion.　　　　True　　False

2. The proportion $\frac{12}{13} = \frac{3}{4}$ is a true proportion.　　　　True　　False

3. In the proportion $\frac{x}{14} = \frac{2}{7}$, $x = 3$ makes a true proportion.　　True　　False

4. In the proportion $\frac{24}{15} = \frac{8}{x}$, $x = 5$ makes a true proportion.　　True　　False

SCALE DRAWINGS

Applying the scale on a map or scale drawing is another way that proportions can be very useful. When the scale is given, and the distance on the map or drawing can be measured, the actual lengths can be calculated. Use a proportion to set the scale measurements equal to the actual measurements.

For example, if the scale on a map is 1 inch = 5 miles, and a school and a house on the map are 12 inches apart, set up a proportion comparing the measures.

The proportion could be set up as:

$$\frac{1 \text{ inch}}{5 \text{ miles}} = \frac{12 \text{ inches}}{x \text{ miles}}$$

Notice how the labels match up to help place the values correctly.

Cross multiply the means and the extremes, and solve the equation.

$$1x = 5 \times 12$$
$$x = 60 \text{ miles}$$

The actual distance between the two locations is 60 miles.

TIP: When you are solving proportion and scale word problems, use the labels to help you line up the values in the correct places. For example, in the previous problem, the labels *inch* and *inches* are lined up across from each other, and the labels *miles* are also lined up across from each other.

PRACTICE 2

1. If the scale on a map is 1 inch = 100 miles, what is the actual distance between two locations 3 inches apart on the map?

2. The scale on a map is $\frac{1}{2}$ inch = 15 miles. How far apart on a map are two cities that are actually located 150 miles apart?

3. The height of a window on a scale drawing is 1.5 inches. If the scale is 1 inch = 4 feet, what is the actual height of the window?

PROPORTIONS WORD PROBLEMS

Proportions are extremely helpful when you are solving certain types of math word problems. By using labels and lining them up in proportion, it can become much easier to set up the correct proportion. Read through the following example to practice these techniques and to set the stage for the next practice section. The four steps to solving math word problems are used for each example.

Example 1

Francis can tile 2 square feet of a floor in 20 minutes. At this rate, how long will it take for him to tile 8 square feet?

Read and understand the question. This question is looking for the total amount of time it will take to tile 8 square feet. The time it took to tile 2 square feet is given.

Make a plan. To solve this problem, use a proportion and line up the corresponding labels to put the values in the correct places.

Carry out the plan. The proportion could be set up as:

$$\frac{2 \text{ square feet}}{20 \text{ minutes}} = \frac{8 \text{ square feet}}{x \text{ minutes}}$$

Notice how the labels match up to help place the values correctly.

Cross multiply the means and the extremes, and solve the equation.

$2x = 20 \times 8$
$2x = 160$
$\frac{2x}{2} = \frac{160}{2}$ Divide each side of the equation by 2.
$x = 80$ minutes

At the same rate, it would take Francis 80 minutes.

Check your answer. Substitute into the proportion to check that it is a true proportion.

$$\frac{2 \text{ square feet}}{20 \text{ minutes}} = \frac{8 \text{ square feet}}{80 \text{ minutes}}$$

The cross products are each 160. This solution is checking.

Example 2

The ratio of boys to girls in an after-school club is 1 to 3. If there are 21 girls in the club, what is the total number of students in the club?

Read and understand the question. This question is looking for the total number of students in a club. The ratio of boys to girls and the number of girls in the club are given.

Make a plan. To solve this problem, use a proportion and line up the corresponding labels to put the values in the correct places. Be aware that the ratio 1 to 3 represents 1 boy for every 3 girls. This question needs a part to whole ratio. Therefore, use the ratio 3 girls for every $1 + 3 = 4$ students.

Carry out the plan. The proportion could be set up as:

$$\frac{3 \text{ girls}}{4 \text{ students}} = \frac{21 \text{ girls}}{x \text{ students}}$$

Cross multiply the means and the extremes, and solve the equation.

$3x = 21 \times 4$
$3x = 84$
$\frac{3x}{3} = \frac{84}{3}$ Divide each side of the equation by 3.
$x = 28$ students

There are a total of 28 students in the club.

Check your answer. Substitute into the proportion to check that it is a true proportion.

$$\frac{3 \text{ girls}}{4 \text{ students}} = \frac{21 \text{ girls}}{28 \text{ students}}$$

The cross products are each 84. This solution is checking.

PRACTICE 3

1. Justine ran the first 6 miles of a 10-mile race in 48 minutes. At this pace, how long will it take her to run the entire 10 miles?

2. Harold makes $204 when he works a shift of 5 hours. How long does he need to work in order to make $326.40?

3. The scale on a drawing is 1 inch = 12 inches. If the length of a wall on the drawing is 9 inches, what is the actual length of the wall?

4. The ratio of flour to salt in a recipe for molding dough is 4:1. If Sam is using 2 cups of salt in his mixture, how many cups of flour should he use?

ANSWERS

Practice 1

1. True; each cross product is equal to 90.
2. False; the cross products are 39 and 48, and they are not equal.
3. False; use $x = 4$ to make it a true proportion.
4. True. When $x = 5$, each of the cross products is 120.

Practice 2

1. The proportion could be set up as:

$$\frac{1 \text{ inch}}{100 \text{ miles}} = \frac{3 \text{ inches}}{x \text{ miles}}$$

Cross multiply the means and the extremes, and solve the equation.

$$1x = 3 \times 100$$

$$x = 300 \text{ miles}$$

The actual distance between the two locations is 300 miles.

2. The proportion could be set up as:

$$\frac{\frac{1}{2} \text{ inch}}{15 \text{ miles}} = \frac{x \text{ inches}}{150 \text{ miles}}$$

Cross multiply the means and the extremes, and solve the equation.

$$15x = \frac{1}{2} \times 150$$

$$15x = 75 \text{ miles}$$

$$\frac{15x}{15} = \frac{75}{15} \qquad \text{Divide each side of the equation by 15.}$$

$$x = 5 \text{ inches}$$

The distance on the map between the two locations is 5 inches.

3. The proportion could be set up as:

$$\frac{1 \text{ inch}}{4 \text{ feet}} = \frac{1.5 \text{ inches}}{x \text{ feet}}$$

Cross multiply the means and the extremes, and solve the equation.

$$1x = 1.5 \times 4$$

$$x = 6 \text{ feet}$$

The actual height of the window is 6 feet.

PRACTICE 3

1. *Read and understand the question.* This question is looking for the total number of minutes it will take to run 10 miles. The time it took to run 6 miles is given.

   Make a plan. To solve this problem, use a proportion and line up the corresponding labels to put the values in the correct places.

   Carry out the plan. The proportion could be set up as:

   $$\frac{6 \text{ miles}}{48 \text{ minutes}} = \frac{10 \text{ miles}}{x \text{ minutes}}$$

   Notice how the labels match up to help place the values correctly.

   Cross multiply the means and the extremes, and solve the equation.

   $$6x = 10 \times 48$$

   $$6x = 480$$

   $$\frac{6x}{6} = \frac{480}{6} \qquad \text{Divide each side of the equation by 6.}$$

   $$x = 80 \text{ minutes}$$

   At the same rate, it would take her 80 minutes.

   Check your answer. Substitute into the proportion to check that it is a true proportion:

   $$\frac{6 \text{ miles}}{48 \text{ minutes}} = \frac{10 \text{ miles}}{80 \text{ minutes}}$$

   The cross products are each 480. This solution is checking.

2. *Read and understand the question.* This question is looking for the total number of hours it will take to make $326.40. The time it took to earn $204 is given.

   Make a plan. To solve this problem, use a proportion and line up the corresponding labels to put the values in the correct places.

   Carry out the plan. The proportion could be set up as:

   $$\frac{\$204}{5 \text{ hours}} = \frac{326.40}{x \text{ hours}}$$

   Line up the labels to help place the values correctly.

   Cross multiply the means and the extremes, and solve the equation.

   $$204x = 5 \times 326.40$$

   $$204x = 1{,}632$$

   $$\frac{204x}{204} = \frac{1{,}632}{204} \qquad \text{Divide each side of the equation by 204.}$$

   $$x = 8 \text{ hours}$$

   At the same rate, it would take Harold 8 hours to make $326.40.

   Check your answer. Substitute into the proportion to check that it is a true proportion:

   $$\frac{\$204}{5 \text{ hours}} = \frac{\$326.40}{8 \text{ hours}}$$

   The cross products are each 1,632. This solution is checking.

3. *Read and understand the question.* This question is looking for the actual length of a wall. The length on a scale drawing is given.

Make a plan. To solve this problem, use a proportion and line up the corresponding labels to put the values in the correct places.

Carry out the plan. The proportion could be set up as:

$$\frac{1 \text{ inch scale}}{12 \text{ inches actual}} = \frac{9 \text{ inches scale}}{x \text{ inches actual}}$$

Notice how the labels match up to help place the values correctly.

Cross multiply the means and the extremes, and solve the equation.

$$1x = 12 \times 9$$
$$x = 108 \text{ inches}$$

The actual length is 108 inches.

Check your answer. Substitute into the proportion to check that it is a true proportion:

$$\frac{1 \text{ inch scale}}{12 \text{ inches actual}} = \frac{9 \text{ inches scale}}{108 \text{ inches actual}}$$

The cross products are each 108. This solution is checking.

4. *Read and understand the question.* This question is looking for the total amount of flour needed. The ratio needed is given in the problem.

Make a plan. To solve this problem, use a proportion and line up the corresponding labels to put the values in the correct places.

Carry out the plan. The proportion could be set up as:

$$\frac{4 \text{ parts flour}}{1 \text{ part salt}} = \frac{x \text{ cups flour}}{2 \text{ cups salt}}$$

Cross multiply the means and the extremes, and solve the equation.

$$1x = 2 \times 4$$
$$x = 8 \text{ cups}$$

A total of 8 cups of flour are needed.

Check your answer. Substitute into the proportion to check that it is a true proportion

$$\frac{4 \text{ parts flour}}{1 \text{ part salt}} = \frac{8 \text{ cups flour}}{2 \text{ cups salt}}$$

Each cross product is 8. This solution is checking.

ⓛ ⓔ ⓢ ⓢ ⓞ ⓝ 11

percent word problems

It is a mathematical fact that fifty percent of all
doctors graduate in the bottom half of their class.
—AUTHOR UNKNOWN

In this lesson, you will review converting among percents, decimals, and fractions, as well as how to apply percents in word problems.

A percent is the ratio of a number compared to 100. Percents are written with the percent symbol. For example, the ratio 12 out of 100 or $\frac{12}{100}$ is written as 12%.

It is helpful to know how to change values among percent form, fraction form, and decimal form.

CHANGING FROM A PERCENT TO A FRACTION

To change a percent to fraction form, write the percent as a number over 100. The percent 79% is equal to $\frac{79}{100}$. The percent 15% is equal to $\frac{15}{100} = \frac{15 \div 5}{100 \div 5} = \frac{3}{20}$ in simplest form.

CHANGING FROM A FRACTION TO A PERCENT

To change a fraction to a percent, set the fraction equal to $\frac{x}{100}$ and cross multiply. The fraction $\frac{4}{25}$ can be changed to a percent by setting up the proportion $\frac{4}{25} = \frac{x}{100}$. Cross multiply to get $25x = 400$. Divide each side of the equation by 25.

$$\frac{25x}{25} = \frac{400}{25}$$
$$x = 16$$

TIP: Another way to change a fraction to a percent is to change the fraction to decimal form, and then multiply by 100. For example, to change $\frac{1}{4}$ to a percent, divide the numerator by the denominator: $1 \div 4 = 0.25$. Next, multiply this decimal by 100 and write the percent symbol after the number: $0.25 \times 100 = 25\%$.

CHANGING A PERCENT TO A DECIMAL

To change a percent to a decimal, divide the percent by 100 and take away the percent symbol. For example, the percent 65% is equal to $65 \div 100 = 0.65$. The percent 3% is equal to $3 \div 100 = 0.03$.

CHANGING A DECIMAL TO A PERCENT

To change a decimal to a percent, simply multiply the decimal by 100 and write the percent symbol after the number. The decimal 0.43 is equal to $0.43 \times 100 = 43\%$. The decimal 2.34 is equal to $2.34 \times 100 = 234\%$.

TIP: Here's a quick way to convert between decimals and percents. When you are changing from a decimal to a percent, move the decimal point two places to the right: $0.45 = 45\%$.

When you are changing from a percent to a decimal, move the decimal point two places to the left: $64\% = 0.64$.

PRACTICE 1

Match the decimal or fraction from column 1 with the correct percent in column 2.

Column 1	Column 2
0.62	15%
$\frac{1}{2}$	40%
0.08	62%
$\frac{9}{100}$	50%
$\frac{2}{5}$	9%
1.15	8%
$\frac{15}{100}$	115%

PERCENT WORD PROBLEMS

Many percent word problems can be solved by using the proportion $\frac{\text{part}}{\text{whole}} = \frac{x}{100}$ or $\frac{\text{is}}{\text{of}} = \frac{x}{100}$. When you are using these proportions, the number in the question with the percent symbol or the word *percent* is placed over 100. The number after the word *of* is the **whole**, or the denominator of the first fraction. The number that is usually closer to the word *is* is the **part**, or the numerator of the first fraction.

Read the following examples to help apply this strategy for certain percent word problems.

Example 1

What percent is 30 of 120?

Read and understand the question. This question is asking for the percent that one number is of another number.

Make a plan. Use the proportion $\frac{\text{is (part)}}{\text{of (whole)}} = \frac{x}{100}$. This question is looking for the percent, so place x over 100. The number after the word *of* is 120, so 120 is the **whole**. The number after the word *is* is 30, so 30 is the **part**.

Carry out the plan. Set up the proportion: $\frac{30}{120} = \frac{x}{100}$. Cross multiply to get $120x$ = 3,000. Divide each side of the equation by 120.

$$\frac{120x}{120} = \frac{3,000}{120}$$
$$x = 25\%$$

Thirty out of 120 is equal to 25%.

Check your answer. Substitute the solution into the proportion to check this answer. The proportion is $\frac{30}{120} = \frac{25}{100}$. The cross products are each 3,000, so this answer is checking.

Example 2

What number is 10% of 80?

Read and understand the question. This question is looking for 10% of a number.

Make a plan. Use the proportion $\frac{\text{is (part)}}{\text{of (whole)}} = \frac{x}{100}$. This question gives the value 10%, so place 10 over 100. The number after the word *of* is 80, so 80 is the **whole**. The phrase *what number* appears before the word *is*, so x is the **part**.

Carry out the plan. Set up the proportion. $\frac{x}{80} = \frac{10}{100}$. Cross multiply to get $100x$ = 800. Divide each side of the equation by 100.

$$\frac{100x}{100} = \frac{800}{100}$$
$$x = 8$$

Eight is 10% of 80.

Check your answer. Substitute the solution into the proportion to check this answer. The proportion is $\frac{8}{80} = \frac{10}{100}$. The cross products are each 800, so this answer is checking.

....

TIP: Another quick way to find the percent of a number is to change the percent to a decimal, and then multiply by the number. For example, to find 20% of 30, change 20% to a decimal and multiply: $0.20 \times 30 = 6$; 20% of 30 is 6.

....

Example 3

54 is 90% of what number?

...

This question is asking for a number

s (part)/(whole) = $\frac{x}{100}$. This question is looking for
of) so place x in the denominator of the
word *is* is 54, so 54 is the **part**, or the
ent given in the question is 90%, so place

ortion. $\frac{54}{x} = \frac{90}{100}$. Cross multiply to get $90x$
n by 90.

tion by finding 90% of 60. This is equal
king.

ANSWERS

Practice 1

Column 1	Column 2
0.62	62%
$\frac{1}{2}$	50%
0.08	8%
$\frac{9}{100}$	9%
$\frac{2}{5}$	40%
1.15	115%
$\frac{15}{100}$	15%

Practice 2

1. *Read and understand the question.* This question is asking for the percent one number is of another number.
   Make a plan. Use the proportion $\frac{\text{is (part)}}{\text{of (whole)}} = \frac{x}{100}$. This question is looking for the percent, so place x over 100. The number after the word *of* is 50, so 50 is the **whole**. The number 40 is the **part**.
   Carry out the plan. Set up the proportion: $\frac{40}{50} = \frac{x}{100}$. Cross multiply to get $50x = 4{,}000$. Divide each side of the equation by 50.
   $$\frac{50x}{50} = \frac{4{,}000}{50}$$
   $$x = 80\%$$
   40 out of 50 is equal to 80%.
   Check your answer. Substitute the solution into the proportion to check this answer. The proportion is $\frac{40}{50} = \frac{80}{100}$. The cross products are each 4,000, so this answer is checking.

2. *Read and understand the question.* This question is looking for the percent of a number.
   Make a plan. Use the proportion $\frac{\text{is (part)}}{\text{of (whole)}} = \frac{x}{100}$. This question gives the value 140%, so place 140 over 100. The number after the word *of* is 20, so 20 is the **whole**. The phrase *what number* appears after the word *is*, so x is the **part**.

Carry out the plan. Set up the proportion. $\frac{x}{20} = \frac{140}{100}$. Cross multiply to get $100x = 2{,}800$. Divide each side of the equation by 100.

$$\frac{100x}{100} = \frac{2{,}800}{100}$$
$$x = 28$$

28 is 140% of 20.

Check your answer. Substitute the solution into the proportion to check this answer. The proportion is $\frac{28}{20} = \frac{140}{100}$. The cross products are each 2,800, so this answer is checking.

3. *Read and understand the question.* This question is asking for a number when the part of the number is given.

   Make a plan. Use the proportion $\frac{\text{is (part)}}{\text{of (whole)}} = \frac{x}{100}$. This question is looking for the **whole** (the number after the word *of*) so place x in the denominator of the first fraction. The number before the word *is* is 10.5, so 10.5 is the **part**, or the numerator of the first fraction. The percent given in the question is 2%, so place 2 over 100 in the proportion.

   Carry out the plan. Set up the proportion. $\frac{10.5}{x} = \frac{2}{100}$. Cross multiply to get $2x = 1{,}050$. Divide each side of the equation by 2.

   $$\frac{2x}{2} = \frac{1{,}050}{2}$$
   $$x = 525$$

   10.5 is 2% of 525.

   Check your answer. Check this solution by finding 2% of 525. This is equal to $0.02 \times 525 = 10.5$, so this problem is checking.

4. *Read and understand the question.* This question is looking for the percent of a number.

   Make a plan. Use the proportion $\frac{\text{is (part)}}{\text{of (whole)}} = \frac{x}{100}$. This question gives the value 45%, so place 45 over 100. The number after the word *of* is 80, so 80 is the **whole**. The phrase *what number* appears before the word *is*, so x is the **part**.

   Carry out the plan. Set up the proportion: $\frac{x}{80} = \frac{45}{100}$. Cross multiply to get $100x = 3{,}600$. Divide each side of the equation by 100.

   $$\frac{100x}{100} = \frac{3{,}600}{100}$$
   $$x = 36$$

   36 is 45% of 80.

   Check your answer. Substitute the solution into the proportion to check this answer. The proportion is $\frac{36}{80} = \frac{45}{100}$. The cross products are each 3,600, so this answer is checking.

applications of percent

*The longer mathematics lives the more abstract—and therefore,
possibly also the more practical—it becomes.*
—ERIC TEMPLE BELL (1883–1960)

This lesson takes the basics of percent and applies them to everyday problems.
You will review important topics such as commission, tax, tip, discounts, and
simple interest, as well as learn how to solve these practical applications of
percent.

APPLICATIONS OF PERCENT WORD PROBLEMS

Most percent word problems can be solved using the proportion $\frac{\text{part}}{\text{whole}} = \frac{x}{100}$. This
important proportion will be adjusted for each of the problem types that follow.

COMMISSION

Commission is the amount of money earned by a salesperson for selling a certain product. These products could be in the form of televisions, appliances, real
estate (homes and property), and automobiles, to name a few. It is common for
a salesperson's commission to be a percent of his or her total sales. To find the
commission earned, use the proportion $\frac{\text{commission } (x)}{\text{total sales (whole)}} = \frac{\%}{100}$.

Example

Charles works at a job selling computer equipment. He receives 4% commission on his total sales for the month. What is the commission he earns if his monthly sales total is $14,560?

To find the commission, use the proportion $\frac{\text{commission } (x)}{\text{total sales (whole)}} = \frac{\%}{100}$. Substitute the values, and the proportion becomes $\frac{x}{14,560} = \frac{4}{100}$. Cross multiply to get $100x = 58,240$. Divide each side of the equation to get

$$\frac{100x}{100} = \frac{58,240}{100}$$
$$x = \$582.40$$

Charles makes $582.40 in commission.

SALES TAX

Sales tax is an amount of money added to a total purchase. This amount is usually a percent of the total sales, and is determined by the county and state in which the purchase is made. Use the proportion $\frac{\text{tax } (x)}{\text{total sales (whole)}} = \frac{\%}{100}$ to solve sales tax problems.

Example

Toby buys a new MP3 player for a price of $45.50. What is the total amount his credit card is charged if the sales tax is 7%?

To find the sales tax, use the proportion $\frac{\text{tax } (x)}{\text{total sales (whole)}} = \frac{\%}{100}$. Substitute the values, and the proportion becomes $\frac{x}{45.50} = \frac{7}{100}$. Cross multiply to get $100x = 318.5$. Divide each side of the equation to get

$$\frac{100x}{100} = \frac{318.5}{100}$$
$$x = 3.185$$

which rounds to $3.19. Toby pays $3.19 in tax. Thus, the total amount that will be charged is $45.50 + $3.19 = $48.69.

DISCOUNT

A discount is an amount subtracted from the price of an item; it is the money you save when buying something. If a discount is a given dollar amount, just subtract this amount from the price of the item. For example, if a shirt is $5.00 off the original price of $20.00, the new price is $20.00 − $5.00 = $15.00.

However, in other cases, the discount is given as a percent off the total amount purchased. Use the proportion $\frac{\text{discount }(x)}{\text{original cost (whole)}} = \frac{\%}{100}$ to find the amount of discount, and then subtract from the original cost to find the sale price.

Example

Tara is shopping for a new jacket. The original price is $55.00 with a discount of 30% off. What is the sale price of the jacket?

To find the sale price, first find the amount of discount. Use the proportion $\frac{\text{discount }(x)}{\text{original cost (whole)}} = \frac{\%}{100}$. Substitute into the proportion and get $\frac{x}{55} = \frac{30}{100}$. Cross multiply to get $100x = 1{,}650$. Divide each side of the equation by 100 to get

$$\frac{100x}{100} = \frac{1{,}650}{100}$$
$$x = \$16.50$$

Tara will save $16.50. Thus, the total sale price will be $55 − $16.50 = $38.50.

· ·

TIP: Tip, also known as gratuity, is the extra amount of money given to a person providing a service. This could be a server at a restaurant, a hairdresser, or a pet groomer, to name a few examples. Tip can be calculated by using the proportion $\frac{\text{tip }(x)}{\text{total amount (whole)}} = \frac{\%}{100}$.

· ·

Example

Yoshi and his family went out to eat at their favorite restaurant. If the total bill was $65.00, what was the amount of the tip, if they left 20% of the total bill for the server?

To find the amount of the tip, use the proportion $\frac{\text{tip }(x)}{\text{total amount (whole)}} = \frac{\%}{100}$. Substitute into the proportion and get $\frac{x}{65} = \frac{20}{100}$. Cross multiply to get $100x = 1{,}300$. Divide each side of the equation to get

$$\frac{100x}{100} = \frac{1{,}300}{100}$$
$$x = \$13.00$$

They left a $13.00 tip for the server.

PERCENT OF CHANGE (INCREASE OR DECREASE)

The percent of change is calculated when an amount is increased or decreased. Use the proportion $\frac{\text{difference in amounts}}{\text{original amount (whole)}} = \frac{x}{100}$ to solve percent change problems.

Percent Increase Example

The height of a plant increased from 1.5 meters to 2.25 meters. What is the percent of increase?
To solve this problem, use the proportion $\frac{\text{difference in amounts}}{\text{original amount (whole)}} = \frac{x}{100}$. Subtract the amounts to find the difference: $2.25 - 1.5 = 0.75$. Substitute into the proportion and get $\frac{0.75}{1.5} = \frac{x}{100}$. Cross multiply to get $1.5x = 75$. Divide each side of the equation by 1.5 to get

$$\frac{1.5x}{1.5} = \frac{75}{1.5}$$
$$x = 50$$

The percent of increase is 50%.

Percent Decrease Example

The cost of a new pair of sneakers was marked down from $49 to $39.20. What was the percent of decrease?
To solve this problem, use the proportion $\frac{\text{difference in amounts}}{\text{original amount (whole)}} = \frac{x}{100}$. Subtract the amounts to find the difference: $\$49 - \$39.20 = \$9.80$. Substitute into the

proportion and get $\frac{9.80}{49} = \frac{x}{100}$. Cross multiply to get $49x = 980$. Divide each side of the equation by 49 to get

$$\frac{49x}{49} = \frac{980}{49}$$
$$x = 20$$

The percent of decrease is 20%.

SIMPLE INTEREST

Simple interest is the amount of interest calculated on a loan or investment. Simple interest is based on the fact that the interest is calculated on an annual basis, or once a year. To find simple interest, use the equation *Interest = Principle × Rate × Time* or $I = p \times r \times t$.

TIP: When using the formula $I = p \times r \times t$, where I is the interest, p is the principal, r is the rate and t is the time, be sure to change the percent that represents the rate to a decimal before substituting into the formula. For example, the rate 8% should be changed to 0.08.

Also, be sure that the time in the formula is in years; if it is given in months, divide by 12 to convert to years. For example, 6 months is equal to $\frac{6}{12} = 0.5$ years.

Example 1

What is the balance on an account that has $500 invested at 4% for 3 years?

In this problem, the principle (p) is $500, the interest rate ($r$) is 4%, and the time ($t$) is 3 years. Substitute each value into the formula. Don't forget to change the percent to a decimal. The formula becomes $I = (500)(0.04)(3) = \$60$. The interest earned is $60.

To find the total balance on the account, add the principle and the interest: $500 + $60 = $560.

Example 2

Suppose $1,400 is invested at an interest rate of 7%. If the interest earned is $392, how long was the money invested?

In this problem, the principle (p) is $1,400, the interest rate (r) is 7%, and the time (t) is unknown. This time, the interest (I) is given as $392. Substitute each value into the formula. Don't forget to change the percent to a decimal. The formula becomes $392 = (1,400)(0.07)(t)$. Multiply on the right side of the equation: $392 = 98t$. Divide each side of the equation by 98.

$$\frac{392}{98} = \frac{98t}{98}$$
$$4 = t$$

The time is 4 years.

PRACTICE

1. Pat makes 3% commission on his total weekly sales. If his weekly sales total $2,350, what is his amount of commission earned?

2. What is the total cost of an item that is marked $20.00 if the sales tax is 8%?

3. What is the sale price of a new computer that costs $450.00 if the amount of discount is 25% off?

4. If a 15% tip was left for a server, what was the amount of the tip in dollars if the total cost of the meal was $24?

5. The price of a gallon of gasoline increased from $2.50 a gallon to $2.75 a gallon. What was the percent of increase?

6. John's electric bill went from $68 last month to $54.40 this month. What was the percent of decrease in his bill?

7. What is the total amount of interest paid on a $150 loan with an 8% interest rate and a time of 2 years?

8. Henry earns $50 interest on money he invested at a 5% interest rate for 6 months. How much money did he invest?

ANSWERS

1. *Read and understand the question.* This question is looking for the amount of commission earned. The percent of commission and total sales are given.

   Make a plan. To find the commission, use the proportion $\frac{\text{commission }(x)}{\text{total sales (whole)}} = \frac{\%}{100}$.

   Carry out the plan. Substitute the values, and the proportion becomes $\frac{x}{2,350} = \frac{3}{100}$. Cross multiply to get $100x = 7,050$. Divide each side of the equation by 100 to get
   $$\frac{100x}{100} = \frac{7,050}{100}$$
   $$x = \$70.50$$
   Pat made \$70.50 in commission.

   Check your answer. To check this problem, divide the answer of \$70.50 by \$2,350 to be sure the commission rate comes out to be 3%: $70.50 \div 2,350 = 0.03 = 3\%$. This solution is checking.

2. *Read and understand the question.* This question is asking for the total cost of an item that is marked \$20.00 if the tax that needs to be added is 8% of this price.

   Make a plan. To find the sales tax, use the proportion $\frac{\text{tax }(x)}{\text{total sales (whole)}} = \frac{\%}{100}$. Then, add this amount to \$20.00 for the final answer.

   Carry out the plan. Substitute the values, and the proportion becomes $\frac{x}{20} = \frac{8}{100}$. Cross multiply to get $100x = 160$. Divide each side of the equation by 100 to get
   $$\frac{100x}{100} = \frac{160}{100}$$
   $$x = \$1.60$$
   The tax is \$1.60. Thus, the total cost is $\$20.00 + \$1.60 = \$21.60$.

   Check your answer. To check this answer, divide the amount of tax by \$20.00 to be sure that the tax rate is 8%: $1.60 \div 20 = 0.08 = 8\%$. This answer is checking.

3. *Read and understand the question.* This question is looking for the sale price when a discount of 25% is given.

   Make a plan. To find the sale price, first find the amount of discount. Use the proportion $\frac{\text{discount }(x)}{\text{original cost (whole)}} = \frac{\%}{100}$. Then, subtract the discount from the original cost to find the sale price.

   Carry out the plan. Substitute into the proportion and get $\frac{x}{450} = \frac{25}{100}$. Cross multiply to get $100x = 11{,}250$. Divide each side of the equation by 100 to get

   $$\frac{100x}{100} = \frac{11{,}250}{100}$$
   $$x = \$112.50$$

   The discount is \$112.50. Thus, the total sale price will be \$450 − \$112.50 = \$337.50.

   Check your answer. To check this answer, divide the amount of discount by 450 to be sure it is 25% of the original price: $112.50 \div 450 = 0.25 = 25\%$. This solution is checking.

4. *Read and understand the question.* This question asks for the amount of tip when the percent is 15%.

   Make a plan. To find the amount of tip, use the proportion $\frac{\text{tip }(x)}{\text{total amount (whole)}} = \frac{\%}{100}$.

   Carry out the plan. Substitute into the proportion and get $\frac{x}{24} = \frac{15}{100}$. Cross multiply to get $100x = 360$. Divide each side of the equation by 100 to get

   $$\frac{100x}{100} = \frac{360}{100}$$
   $$x = \$3.60$$

   They should leave \$3.60 as a tip for the server.

   Check your answer. To check this answer, divide the amount of the tip by \$24 to be sure it is 15% of the original price: $\$3.60 \div \$24 = 0.15 = 15\%$. This answer is checking.

5. *Read and understand the question.* This question is looking for the percent of increase in the cost of a gallon of gasoline.

   Make a plan. To solve this problem, use the proportion $\frac{\text{difference in amounts}}{\text{original amount (whole)}} = \frac{x}{100}$. Subtract the two given amounts to find the difference.

   Carry out the plan. Subtract the amounts to find the difference: \$2.75 − \$2.50 = \$0.25. Substitute into the proportion and get $\frac{0.25}{2.5} = \frac{x}{100}$. Cross multiply to get $2.5x = 25$. Divide each side of the equation by 2.5 to get

   $$\frac{2.5x}{2.5} = \frac{25}{2.5}$$
   $$x = 10$$

The percent of increase is 10%.

Check your answer. Substitute the solution into the proportion to check this answer. The proportion is $\frac{0.25}{2.50} = \frac{10}{100}$. The cross products are each 25, so this answer is checking.

6. *Read and understand the question.* This question asks for the percent of decrease between John's electric bills.

   Make a plan. To solve this problem, use the proportion $\frac{\text{difference in amounts}}{\text{original amount (whole)}} = \frac{x}{100}$. Subtract the two given amounts to find the difference.

   Carry out the plan. Subtract the amounts to find the difference: $68 - \$54.40 = \$13.60$. Substitute into the proportion and get $\frac{13.60}{68} = \frac{x}{100}$. Cross multiply to get $68x = 1,360$. Divide each side of the equation by 68 to get

   $$\frac{68x}{68} = \frac{1,360}{68}$$
   $$x = 20$$

   The percent of decrease is 20%.

   Check your answer. Substitute the solution into the proportion to check this answer. The proportion is $\frac{13.60}{68} = \frac{20}{100}$. The cross products are each 1,360, so this answer is checking.

7. *Read and understand the question.* This question is asking for the amount of interest on a loan after 2 years.

   Make a plan. Use the formula $I = p \times r \times t$, and substitute the given values.

   Carry out the plan. In this problem, the principle (p) is $150, the interest rate ($r$) is 8%, and the time ($t$) is 2 years. Substitute each value into the formula. Don't forget to change the percent to a decimal. The formula becomes $I = (150)(0.08)(2) = \$24$. The interest earned is $24.

   Check your answer. To check this answer, divide the amount of interest by the principle and the rate to see if the result is the time of 2 years: $\frac{24}{(150)(0.08)} = 2$ years. This question is checking.

8. *Read and understand the question.* This question is asking for the amount of principle on an investment of 6 months.

Make a plan. Use the formula $I = p \times r \times t$ and substitute the given values.

Carry out the plan. In this problem, the principle (p) is unknown, the interest rate (r) is 5%, and the time (t) is 6 months, or 0.5 years. This time, the interest (I) is given as $50. Substitute each value into the formula. Don't forget to change the percent to a decimal. The formula becomes $50 = (p)(0.05)(0.5)$. Multiply on the left side of the equation: $50 = 0.025p$. Divide each side of the equation by 0.025.

$$\frac{50}{0.025} = \frac{0.025p}{0.025}$$

$$2{,}000 = p$$

The principle is $2,000.

Check your answer. To check this answer, divide the amount of interest by the principle and the rate to see if the result is the time of 0.5 years: $\frac{50}{(2{,}000)(0.050)} = 0.5$ years. This answer is checking.

S E C T I O N 3

algebra word problems— finding the unknown in unknown territory

ALGEBRA OFTEN PRESENTS its own challenges to the solver, and algebra word problems are no exception. However, you can make these problems easier to solve by applying many of the strategies already studied in this book. The idea of translating from word phrases into numbers and symbols can be used when you are writing a formula to solve certain algebra word problems. In addition, you can solve other types of algebra questions by looking for a pattern, or solving a simpler problem, so do not forget to try these strategies and others if you get stuck. Keep a positive attitude; the road to mastering algebra word problems may appear as a bumpy trail in unknown territory, but you can pave it into a smooth road with the tools mentioned in this section and throughout the book.

This section specifically discusses algebra word problems including equation solving strategies and common types of word problems involving algebra.

This section will introduce you to algebra word problems including:

- translating expressions
- translating equations
- solving equations
- working with inequalities
- applying the rules of exponents
- applications of algebra

simplifying expressions and solving equation word problems

Mathematics may be defined as the economy of counting.
There is no problem in the whole of mathematics
which cannot be solved by direct counting.
—ERNST MACH (1838–1916)

This lesson reviews the key words and phrases for the basic operations and provides examples and tips on simplifying algebraic expressions and the equation solving steps. Equation word problems are modeled with explanations to help your understanding in this type of question.

KEY WORDS AND PHRASES

Translating expressions from words into mathematical symbols was covered in Lesson 1. The following chart below summarizes the key words and phrases studied in that lesson for the four basic operations and the equal sign.

+	−	×	÷	=
sum	difference	product	quotient	is
increased	decreased	times	divide	total
combine	less than	factor	into	result
plus	take away	twice (2×'s)	out of	same as
more than		triple (3×'s)	split	equivalent to
			break up	

Refer to this chart when you are changing sentences in words to math equations.

TIP: In algebra, the number in front of the letter is called the coefficient and the letter is called the variable. In the expression $8x$, 8 is the coefficient and x is the variable.

SIMPLIFYING EXPRESSIONS

Two important processes to know when you are simplifying expressions are combining like terms and the distributive property.

COMBINING LIKE TERMS

Terms, in mathematics, are numbers and symbols that are separated by addition and subtraction.

The expressions 3, $5x$, and $7xy$ are each one term.

The expressions $2x + 3$, and $x − 7$ each have two terms.

The expression $7x + 5y − 9$ has three terms.

Like terms are terms with the same variable and exponent. Like terms can be combined by addition and subtraction. To do this, add or subtract the coefficients and keep the variable the same. For example $3x + 5x = 8x$, and $6y^2 − 4y^2 = 2y^2$.

TIP: Be sure to combine only **like terms**. Terms without the exact same variable and exponent cannot be combined: $5x^2$ and $6x$ cannot be combined because the exponents are not the same.

DISTRIBUTIVE PROPERTY

The distributive property is used when a value needs to be multiplied, or distributed, to more than one term. For example, in the expression $3(x + 10)$, the number 3 needs to be multiplied by the term x and the term 10. The use of arrows can help in this process, as shown in the following figure.

$$3\ (x = 10)$$

The result becomes $3 \times x + 3 \times 10$, which simplifies to $3x + 30$.

PRACTICE 1

Simplify each of the following expressions:

1. $4x + 10x = $ _____

2. $20y - 3y = $ _____

3. $-9x^2 + 6x^2 - 2x = $ _____

4. $2(x - 4) = $ _____

5. $7(2y + 5) = $ _____

SOLVING EQUATIONS

When you are solving equations, the goal is to get the letter, or variable, by itself. This is called isolating the variable. Each of the following examples goes through the process of isolating the variable for different types of equations.

TIP: One of the most important rules in equation solving is to **do the same thing on both sides** of the equation. For example, if you divide on one side to get the variable alone, divide the other side by the same number. This keeps the equation balanced and will lead to the correct solution.

ONE-STEP EQUATIONS

One-step equations are named this because they have only one operation to *undo*, so they should take only one step to solve. Use inverse, or opposite, operations to get the variable alone. This is called *isolating the variable*.

Example: Solve the equation for x.

$3x = 12$

$\frac{3x}{3} = \frac{12}{3}$ Divide each side by 3, the inverse operation of multiplying by 3.

$x = 4$ The variable is alone.

TWO-STEP EQUATIONS

Two-step equations have two operations, and therefore take two steps to solve.

Example: Solve the equation for x.

$\frac{x}{4} - 10 = 10$

$\frac{x}{4} - 10 + 10 = 10 + 10$ Add 10 to each side, the inverse operation of subtracting 10.

$\frac{x}{4} = 20$ Simplify.

$\frac{x}{4} \times 4 = 20 \times 4$ Multiply each side by 4, the inverse operation of dividing by 4.

$x = 80$ The variable is alone.

VARIABLES ON BOTH SIDES OF THE EQUATION

When there are variables on both sides of the equation, get rid of the variable with the smaller coefficient by using the inverse operation.

Example: Solve the equation for x.

$7x = 3x + 28$

$7x - 3x = 3x - 3x + 28$ Subtract $3x$ from each side of the equation.

$4x = 28$ Simplify.

$\frac{4x}{4} = \frac{28}{4}$ Divide each side by 4, the inverse operation of multiplying by 4.

$x = 7$ The variable is alone.

MULTISTEP EQUATIONS

To solve multistep equations, you may first need to use the distributive property and combine like terms in order to simplify. If there are variables on both sides of the equation, handle them next. Finally, use the inverse operations to isolate the variable in the one- or two-step equation.

Example: Solve the equation for x.

$2(x - 6) - x = 5x$

$2x - 12 - x = 5x$ Get rid of the parentheses by using the distributive property.

$x - 12 = 5x$ Combine like terms.

$x - x - 12 = 5x - x$ Subtract x from each side of the equation.

$-12 = 4x$ Simplify.

$\frac{-12}{4} = \frac{4x}{4}$ Divide each side by 4, the inverse operation of multiplying by 4.

$-3 = x$ The variable is alone.

TIP: The equation solving steps can be summarized as the following:
1. Get rid of parentheses.
2. Combine like terms.
3. Handle variables of both sides of the equation.
4. Solve the one- or two-step equation.

PRACTICE 2

Solve each of the following equations.

1. $x + 10 = 25$

2. $3x - 5 = 10$

3. $12x - 3x = 18$

4. $5(x + 1) = 30$

5. $2x + 10 = 3(x - 4)$

EQUATION WORD PROBLEMS

Now that the steps to solving equations have been practiced, let's apply these steps to solving word problems involving equations. Use the chart at the beginning of the lesson and the examples in Lesson 1 for help with translating phrases into math symbols and equations. Then, use your knowledge and skills in equation solving to find the correct solution to each problem. In addition, use the word-problem solving steps to be sure each detail is taken care of and all problems are checked.

TIP: To check solutions in equations, substitute the value into the original equation and use **order of operations**. The correct order of operations is Parentheses, Exponents, Multiplication and Division, Addition and Subtraction. It is commonly remembered as the acronym **PEMDAS**.

Example 1

Ten more than a number is equal to 40. What is the number?

Read and understand the question. This question is looking for a number when clues about this number are given.

Make a plan. Translate the statement into equation form. Then, solve the equation using the equation solving steps.

Carry out the plan. Let x = a number. The key phrase *more than* means addition. The statement translates to $x + 10 = 40$. Next, subtract 10 from each side of the equation to get the variable alone.

$x + 10 - 10 = 40 - 10$
$x = 30$

Check your answer. Check your solution by substituting the answer into the equation.

$x + 10 = 40$

becomes

$30 + 10 = 40$
$40 = 40$

This answer is checking.

Example 2

Eight less than twice a number is equal to four times the number. What is the number?

Read and understand the question. This question is looking for a number when clues about this number are given.

Make a plan. Translate the statement into equation form. Then, solve the equation using the equation solving steps.

Carry out the plan. Let x = a number. The key phrase *less than* means subtraction and *twice a number* is written as $2x$. The first part of the statement translates to $2x - 8$. In the second part of the sentence, *four times the number* is written as $4x$. The entire equation is

$$2x - 8 = 4x$$

Get the variables on one side of the equation by subtracting $2x$ from each side.

$$2x - 2x - 8 = 4x - 2x$$

The equation simplifies to

$$-8 = 2x$$

Next, divide each side by 2 to get the variable alone.

$$\frac{-8x}{2} = \frac{2x}{2}$$
$$x = -4.$$

Check your answer. Check your solution by substituting the answer into the equation.

$$2x - 8 = 4x$$

becomes

$$2(-4) - 8 = 4(-4)$$
$$-8 - 8 = -16$$
$$-16 = -16$$

This answer is checking.

Example 3

Forty-two added to a number is equal to 6 times the sum of the number and 2. What is the number?

Read and understand the question. This question is looking for a number when clues about this number are given.

Make a plan. Translate the statement into equation form. Then, solve the equation using the equation solving steps.

Carry out the plan. Let x = a number. The key phrase *added to* means addition. The first part of the statement translates to $x + 42$. The second part of the sentence, *six times the sum of a number and 2* is written as $6(x + 2)$. The entire equation is

$$x + 42 = 6(x + 2)$$

Use the distributive property on the right side to make the equation

$$x + 42 = 6x + 12$$

Get the variables on one side of the equation by subtracting x from each side.

$$x - x + 42 = 6x - x + 12$$

The equation simplifies to

$$42 = 5x + 12$$

Subtract 12 from each side of the equation to get $30 = 5x$. Next, divide each side by 5 to get the variable alone:

$$\frac{30}{5} = \frac{5x}{5}$$
$$x = 6$$

Check your answer. Check your solution by substituting the answer into the equation.

$$x + 42 = 6(x + 2)$$

becomes

$$6 + 42 = 6(6 + 2)$$
$$48 = 6(8)$$
$$48 = 48$$

This answer is checking.

PRACTICE 3

1. Six times a number is equal to 300. What is the number?

2. A number decreased by 7 is equal to the product of 4 and 10. What is the number?

3. The sum of five and a number is equal to twice the number. What is the number?

4. Three times the sum of a number and 1 is equal to 21. What is the number?

5. Thirty-one minus a number is the same as twice a number plus 10. What is the number?

ANSWERS

PRACTICE 1

1. $14x$
2. $17y$
3. $-3x^2 - 2x$
4. $2x - 8$
5. $14y + 35$

PRACTICE 2

1. $x + 10 - 10 = 25 - 10$
   $x = 15$
2. $3x - 5 + 5 = 10 + 5$
   $3x = 15$
   $x = 5$
3. $12x - 3x = 18$
   $9x = 18$
   $x = 2$

4. $5(x + 1) = 30$

$5x + 5 = 30$

$5x + 5 - 5 = 30 - 5$

$5x = 25$

$x = 5$

5. $2x + 10 = 3(x - 4)$

$2x + 10 = 3x - 12$

$2x - 2x + 10 = 3x - 2x - 12$

$10 = x - 12$

$10 + 12 = x - 12 + 12$

$x = 22$

PRACTICE 3

1. *Read and understand the question.* This question is looking for a number when clues about this number are given.

Make a plan. Translate the statement into equation form. Then, solve the equation using the equation solving steps.

Carry out the plan. Let x = a number. The key phrase *six times a number* is written as $6x$, so the equation is $6x = 300$. Divide each side by 6 to get the variable alone.

$\frac{6x}{6} = \frac{300}{6}$

$x = 50$

Check your answer. Check your solution by substituting the answer into the equation.

$6x = 300$

becomes

$6(50) = 300$

$300 = 300$

This answer is checking.

2. *Read and understand the question.* This question is looking for a number when clues about this number are given.

Make a plan. Translate the statement into equation form. Then, solve the equation using the equation solving steps.

Carry out the plan. Let x = a number. The key phrase *decreased by* means subtraction, so the first part of the statement translates to $x - 7$. In the second part of the sentence, the key word *product* means multiplication, so multiply 4 by 10 to get 40. The entire equation is $x - 7 = 40$. Get the variable alone by adding 7 to each side. The equation simplifies to $x = 47$.

Check your answer. Check your solution by substituting the answer into the equation.

$$x - 7 = 4 \times 10$$

becomes

$$47 - 7 = 40$$
$$40 = 40$$

This answer is checking.

3. *Read and understand the question.* This question is looking for a number when clues about this number are given.

Make a plan. Translate the statement into equation form. Then, solve the equation using the equation solving steps.

Carry out the plan. Let x = a number. The key word *sum* means addition, so the first part of the statement translates to $x + 5$. In the second part of the sentence, *twice the number* is written as $2x$. The entire equation is $x + 5 = 2x$. Get the variables on one side of the equation by subtracting x from each side.

$$x - x + 5 = 2x - x$$

The equation simplifies to

$$5 = x$$

Check your answer. Check your solution by substituting the answer into the equation.

$$x + 5 = 2x$$

becomes

$$5 + 5 = 2(5)$$
$$10 = 10$$

This answer is checking.

4. *Read and understand the question.* This question is looking for a number when clues about this number are given.

Make a plan. Translate the statement into equation form. Then, solve the equation using the equation solving steps.

Carry out the plan. Let x = a number. The key word *sum* means addition, so the sum of a number and 1 is written as $x + 1$. Multiply this expression by 3 and set it equal to 21. The statement translates to

$$3(x + 1) = 21$$

Apply the distributive property.

$$3x + 3 = 21$$

Subtract 3 from each side of the equation to get $3x = 18$. Next, divide each side by 3 to get the variable alone.

$$\frac{3x}{3} = \frac{18}{3}$$
$$x = 6$$

Check your answer. Check your solution by substituting the answer into the equation.

$$3(x + 1) = 21$$

becomes

$$3(6 + 1) = 21$$
$$3(7) = 21$$
$$21 = 21$$

This answer is checking.

5. *Read and understand the question.* This question is looking for a number when clues about this number are given.

Make a plan. Translate the statement into equation form. Then, solve the equation using the equation solving steps.

Carry out the plan. Let $x =$ a number. The first part of the statement translates to $31 - x$. In the second part of the sentence, *twice a number plus 10* is written as $2x + 10$. The entire equation is

$$31 - x = 2x + 10$$

Get the variables on one side of the equation by adding x to each side.

$$31 - x + x = 2x + x + 10$$

The equation simplifies to $31 = 3x + 10$

Subtract 10 from each side of the equation to get $21 = 3x$. Next, divide each side by 3 to get the variable alone.

$$\frac{21}{3} = \frac{3x}{3}$$
$$x = 7$$

Check your answer. Check your solution by substituting the answer into the equation.

$$31 - x = 2x + 10$$

becomes

$$31 - 7 = 2(7) + 10$$
$$24 = 14 + 10$$
$$14 = 14$$

This answer is checking.

inequality word problems

Still more astonishing is that world of
rigorous fantasy we call mathematics.
—GREGORY BATESON (1904–1980)

This lesson details the basics of translating and solving inequalities and word problems with inequalities. Use the information in Lesson 1 for additional review of the basics of translating words into symbols.

TRANSLATING FROM WORDS TO INEQUALITIES

The key words and phrases are summarized in the following chart.

>	≥	<	≤
is more than	minimum	is smaller than	maximum
is greater than	at least	is less than	at most
is larger than	is not less than	below	not more than
above	not smaller than		is not greater than

Read through the following examples for more help with translating inequalities. Let x = a number in each example. The key phrases are in italics.

Examples

1. Two *is less than* a number $2 < x$
2. Four more than a number *is greater than or equal to* five. $x + 4 \geq 5$
3. The *maximum* value of a number is 28. $x \leq 28$
4. The sum of a number and nine *is at least* 81. $x + 9 \geq 81$

PRACTICE 1

Using the list of phrases and key words from the chart, translate each of the following into mathematical symbols. Let x = a number in each exercise.

1. Twenty is more than a number.   _____

2. The difference of a number and three is not
   more than two.   _____

3. The minimum value of a number is 65.   _____

4. Thirty times a number is at most 90.   _____

SOLVING INEQUALITIES

Solve inequalities by using the same golden rule as equation solving: Whatever you do to one side of the inequality, do to the other side.

Example 1

Solve the inequality for x:

$$4x > 16$$
$$\frac{4x}{4} > \frac{16}{4} \qquad \text{Divide each side by 4.}$$
$$x > 4$$

TIP: When you are dividing or multiplying each side of an inequality by a negative value, the inequality symbol must switch directions. For example, to solve the inequality $-5x > 10$, divide each side by -5. The inequality becomes $x < -2$.

Example 2:

Solve the inequality for x:

$\frac{x}{-2} + 4 \leq 8$

$\frac{x}{-2} + 4 - 4 \leq 8 - 4$ Subtract 4 from each side.

$\frac{x}{-2} \leq 4$ Simplify.

$\frac{x}{-2} \times -2 \geq 4 \times -2$ Multiply each side by -2. Switch the inequality symbol.

$x = -8$ Simplify.

PRACTICE 2

Solve each inequality for x.

1. $x - 5 > 23$

2. $\frac{x}{6} < -9$

3. $-3x + 4 \leq 16$

4. $5(x - 2) \geq 7x$

INEQUALITY WORD PROBLEMS

To solve inequality word problems, first translate from words into mathematical symbols. Then, solve the inequality using the same rules as equation solving with two differences. The solution will not be a single value, but usually a set of values. In addition, if you multiplied or divided each side of the equation by a negative number, you must switch the inequality symbol to the other direction.

Example 1

Five more than twice a number is at least 45. What is the minimum value of the number?

Read and understand the question. This question is looking for a number when clues about this number are given.

Make a plan. Translate the statement into inequality form. Then, solve for x using the inequality solving steps.

Carry out the plan. Let $x =$ a number. The key phrase *more than* means addition, and twice a number is written as $2x$. The first part of the statement translates to $2x + 5$. The key phrase *is at least* means greater than or equal to. The entire inequality is $2x + 5 \geq 45$. Subtract 5 from each side of the inequality to get $2x \geq 40$. Next, divide each side by 2 to get the variable alone.

$$\frac{2x}{2} \geq \frac{40}{2}$$
$$x = 20$$

Check your answer. Check your solution by substituting the answer into the inequality.

$$2x + 5 = 45$$

becomes

$$2(20) + 5 = 45$$
$$40 + 5 = 45$$
$$45 = 45$$

This statement is true, so the solution is checking.

Example 2

There are dogs and cats in a kennel. The number of cats is twice the number of dogs. What is the greatest number of cats in the kennel if there are at most a total of 48 animals in the kennel?

Read and understand the question. This question is looking for a number of cats in the kennel. There are at most a total of 48 animals, so there are 48 animals or less in the kennel.

Make a plan. Translate the statement into inequality form using the clues. Then, solve for x using the inequality solving steps.

Carry out the plan. Let x = the number of dogs. Therefore, $2x$ = the number of cats. Because the number of animals is at most 48, add the number of cats and dogs together and set it *less than or equal to* 48. The inequality is $x + 2x \le 48$. Combine like terms to get $3x \le 48$. Next, divide each side by 3 to get the variable alone.

$$\frac{3x}{3} \le \frac{48}{3}$$
$$x = 16$$

Thus, the number of cats is at most $2(16) = 32$.

Check your answer. Check your solution by adding the number of cats and dogs: $16 + 32 = 48$, which was the most the number of animals could be. This solution is checking.

PRACTICE 3

1. Four more than a number is at least 41. What is the least value of the number?

2. Twice a number decreased by 7 is less than or equal to 21. What is the greatest value of the number?

3. A number divided by –3 is at most –10. What is the minimum value of the number?

4. Tyler and Melissa each have gumdrops. The amount that Tyler has is equal to five more than the amount that Melissa has. What is the minimum number of gumdrops that Tyler has if there are at least 31 gumdrops between them?

ANSWERS

PRACTICE 1

1. $20 > x$
2. $x - 3 \leq 2$
3. $x \geq 65$
4. $30x \leq 90$

PRACTICE 2

1. Add 5 to each side of the inequality: $x > 28$.
2. Multiply each side of the inequality by 6: $x < -54$.
3. Subtract 4 from each side of the inequality to get $-3x = 12$. Divide each side by -3 and switch the inequality symbol: $x = -4$.
4. Use the distributive property on the left side of the inequality to get $5x - 10 = 7x$. Subtract $5x$ from each side to get $-10 = 2x$. Divide each side of the inequality by 2: $-5 = x$.

PRACTICE 3

1. *Read and understand the question.* This question is looking for a number when clues about this number are given.
   Make a plan. Translate the statement into inequality form. Then, solve for x using the inequality solving steps.
   Carry out the plan. Let x = a number. The key phrase *more than* means addition. The first part of the statement translates to $x + 4$. The key phrase *is at least* means is greater than or is equal to. The entire inequality is $x + 4 = 41$. Subtract 4 from each side of the inequality to get $x = 37$, so 37 is the least value of the number.
   Check your answer. Check your solution by substituting the answer into the inequality.
   $$x + 4 = 41$$
   becomes
   $$37 + 4 = 41$$
   $$41 = 41$$
   This statement is true, so the solution is checking.

2. *Read and understand the question.* This question is looking for a number when clues about this number are given.

   Make a plan. Translate the statement into inequality form. Then, solve for x using the inequality solving steps.

   Carry out the plan. Let $x =$ a number. The key phrase *decreased by* means subtraction, and twice a number is written as $2x$. The first part of the statement translates to $2x - 7$. This amount is less than or equal to 21, so the entire inequality is $2x - 7 = 21$. Add 7 to each side of the inequality to get $2x = 28$. Next, divide each side by 2 to get the variable alone.

   $$\frac{2x}{2} \leq \frac{28}{2}$$
   $$x = 14$$

   The greatest value of the number is 14.

   Check your answer. Check your solution by substituting the answer into the inequality.

   $$2x - 7 = 21$$

   becomes

   $$2(14) - 7 = 21$$
   $$28 - 7 = 21$$
   $$21 = 21$$

   This statement is true, so the solution is checking.

3. *Read and understand the question.* This question is looking for a number when clues about this number are given.

   Make a plan. Translate the statement into inequality form. Then, solve for x using the inequality solving steps.

   Carry out the plan. Let $x =$ a number. The key phrase *is at most* means less than or equal to, so the statement translates to the inequality $\frac{x}{-3} \leq -10$. Multiply each side by -3 and switch the direction of the inequality symbol.

   $$-3 \times \frac{x}{-3} \geq -10 \times -3$$
   $$x = 30$$

   The minimum value is 30.

   Check your answer. Check your solution by substituting the answer into the inequality.

   $$\frac{x}{-3} \leq -10$$

   becomes

   $$\frac{30}{-3} \leq -10$$
   $$-10 = -10$$

   This statement is true, so the solution is checking.

4. *Read and understand the question.* This question is looking for the minimum number of gumdrops Tyler has. Tyler has five more than Melissa and there are 31 or more gumdrops between them.

Make a plan. Translate the statement into inequality form. Then, solve for x using the inequality solving steps.

Carry out the plan. Let x = the number of gumdrops Melissa has, and let $x + 5$ = the number of gumdrops Tyler has. The key phrase *is at least* means is greater than or equal to, so add the amounts each person has to get the inequality $x + x + 5 = 31$. Combine like terms to simplify the inequality to $2x + 5 = 31$. Subtract 5 from each side of the inequality to get $2x = 26$. Next, divide each side by 2 to get the variable alone.

$$\frac{2x}{2} \geq \frac{26}{2}$$
$$x = 13$$

Melissa has at least 13 gumdrops, and Tyler has at least $13 + 5 = 18$ gumdrops.

Check your answer. Check your solution by adding the amounts each person has, to be sure this amount is at least 31: $13 + 18 = 31$, so the solution is checking.

working with exponents

Infinity is a floorless room without walls or ceilings.
—AUTHOR UNKNOWN

This lesson will explain the rules for exponents and the various forms of numbers. Scientific notation examples and word problems involving exponents will also be demonstrated.

THE BASICS OF EXPONENTS

Exponents are a way of writing large and small numbers in a shortened way. In the expression 2^3, the number 2 is the base and the number 3 is the exponent. The exponent tells how many times the base is used as a factor, or how many times it should be multiplied. For example, the expression 2 to the third power is written as $2^3 = 2 \times 2 \times 2 = 8$. 2^3 is the exponential form, $2 \times 2 \times 2$ is the expanded form, and 8 is the standard form of this number.

....

TIP: There are three useful forms of numbers when working with exponents. They are:

Exponential form: 2^3
Expanded form: $2 \times 2 \times 2$
Standard form: 8

....

EXPONENT RULES

The rules of exponents can be explained by using the different forms of numbers.

Multiplying

The expressions 4^5 and 4^2 have the same base. In expanded form, $4^5 = 4 \times 4 \times 4 \times 4 \times 4$ and $4^2 = 4 \times 4$. To multiply them together, the expression becomes $4^5 \times 4^2 = 4 \times 4 \times 4 \times 4 \times 4 \times 4 \times 4$. This expanded form is multiplying seven 4s together; in other words, 4^7. This pattern can be explained as $4^5 \times 4^2 = 4^{5+2} = 4^7$. When multiplying with like bases, add the exponents.

Dividing

To divide like bases, such as $3^4 \div 3^3$, use expanded form and write division as a fraction to find a pattern. The expression becomes $\frac{3^4}{3^3} = \frac{\cancel{3} \times \cancel{3} \times \cancel{3} \times 3}{\cancel{3} \times \cancel{3} \times \cancel{3}} = 3^1 = 3$. Notice how the common factors in the numerator and denominator cancel. This pattern can be summarized as $3^4 \div 3^3 = 3^{4-3} = 3^1 = 3$. When dividing like bases, subtract the exponents.

Raising a Power to Another Power

When you are raising an exponent to a power, such as $(2^4)^3$, the base becomes the expression within the parentheses. Therefore, to simplify this expression, write the base 2^4 as a factor three times. This is $2^4 \times 2^4 \times 2^4 = 2^{4+4+4} = 2^{12}$. When raising a power to another power, multiply the exponents.

....

TIP: The rules for exponents can be summarized as the following:

When multiplying like bases, **add the exponents.**
$$x^3 \times x^4 = x^{3+4} = x^7$$
When dividing like bases, **subtract the exponents.**
$$\frac{x^6}{x^2} = x^{6-2} = x^4$$
When raising a power to another power, **multiply the exponents.**
$$(x^3)^2 = x^{3 \times 2} = x^6$$

When adding or subtracting expressions with exponents, follow the **order of operations.**

$$3^2 - 2^3 = 9 - 8 = 1$$

NEGATIVE EXPONENTS

A negative exponent tells you to take the reciprocal of the base. For example, $3^{-2} = \frac{1}{3^2} = \frac{1}{9}$. For expressions to be simplified, they should be written with positive exponents.

PRACTICE 1

Simplify each of the following expressions.

1. $2^3 \times 2^5 =$

2. $2^3 \times 3^2 =$

3. $x^7 \times x^6 =$

4. $\dfrac{5^4}{5^3} =$

5. $\dfrac{x^3}{x^7} =$

6. $7^{-2} =$

SCIENTIFIC NOTATION

Scientific notation is used to write very large and very small numbers in a more efficient manner. This form uses a number between one and ten and multiplies it by a power of 10. For example, the number 3,000,000,000 is written as 3×10^9 in scientific notation.

CONVERTING FROM STANDARD FORM TO SCIENTIFIC NOTATION

To change a number from standard form into scientific notation, take the first non-zero digit and place a decimal point to its right to form a value between 1 and 10. Then, multiply by a power of 10, where the exponent is the number of places the decimal point moves.

For example, to write the number 4,500,000 in scientific notation, use the digits 4 and 5 and write the decimal 4.5. Because the decimal point needs to move six places to the left to be between the 4 and 5, the exponent is 6. The scientific notation is 4.5×10^6.

CONVERTING FROM SCIENTIFIC NOTATION TO STANDARD FORM

To convert to standard form from scientific notation, reverse the process just explained. Write the value between one and 10, but move the decimal point as according to the exponent.

For example, for the value 3.42×10^{-4}, first write the number 3.42. Then, move the decimal to the right if the exponent is positive or to the left if the exponent is negative. Use zeros as placeholders when necessary. Since this exponent is –4, move the decimal four places to the left. The standard form is 0.000342.

TIP: When you are converting from standard form to scientific notation, numbers greater than one have a positive exponent and numbers less than one have a negative exponent.

PRACTICE 2

Match the scientific notation with the correct standard form

1. 4.3×10^{-3} a. 43

2. 4.3×10^3 b. 0.0000043

3. 4.3×10^{-6} c. 4,300

4. 4.3×10^{-1} **d.** 0.0043

5. 4.3×10^{1} **e.** 0.43

WORD PROBLEMS WITH EXPONENTS

The word problems in this section involve using the properties of exponents. Use the information in this lesson and the problem solving steps to solve each one.

Example 1

What is the product of $10^5 \times 10^8$?

Read and understand the question. This question is looking for the product of two values that have the same base.

Make a plan. Add the exponents to find the solution.

Carry out the plan. The problem becomes $10^5 \times 10^8 = 10^{5+8} = 10^{13}$.

Check your answer. To check your answer, divide the product by one of the factors that was multiplied in the question.

$$\frac{10^{13}}{10^5} = 10^{13-5} = 10^8$$

Because this was the other factor, this answer is checking.

Example 2

What is the quotient of 16^5 and 16^3?

Read and understand the question. This question is looking for the quotient of two values that have the same base.

Make a plan. Subtract the exponents to find the solution.

Carry out the plan. The problem becomes

$$\frac{16^5}{16^3} = 16^{5-3}$$

$$= 16^2$$

Check your answer. To check your answer, multiply the quotient by the factor 16^3.

$$16^2 \times 16^3 = 16^{2+3}$$
$$= 16^5$$

Because this was the first value given in the question, this answer is checking.

Example 3

A garbage company collects 64,000,000 pounds of garbage per year. What is this amount expressed in scientific notation?

Read and understand the question. This question is looking for the scientific notation when the standard form is given.

Make a plan. Use the first non-zero digit and a decimal point to form a number between one and 10, and then write this number multiplied by a power of 10. The exponent is the number of places the decimal point moves from the end of the number to between the 6 and the 4.

Carry out the plan. Write the decimal 6.4. Next, multiply this value by 10^7 because the decimal moves 7 places to the left. The scientific notation is 6.4×10^7.

Check your answer. To check your answer, put this number back into standard form. Take the decimal 6.4 and move the decimal 7 places to the right. Add zeros as placeholders where necessary. The number becomes 64,000,000. This answer is checking.

PRACTICE 3

1. What is the product of 4^3 and 4^5?

2. What is the quotient of $\dfrac{x^{10}}{x^3}$?

3. What is the product of y^{12} and y^4?

4. What is the quotient of $7^5 \div 7^8$?

5. The population of a country is written as 4.23×10^6 people. What is the population expressed in standard form?

6. The diameter of a microscopic cell is 0.00056 cm. What is the diameter expressed in scientific notation?

ANSWERS

PRACTICE 1

1. $2^3 \times 2^5 = 2^{3+5} = 2^8$
2. $2^3 \times 3^2 = 2 \times 2 \times 2 \times 3 \times 3 = 8 \times 9 = 72$
3. $x^7 \times x^6 = x^{7+6} = x^{13}$
4. $\frac{5^4}{5^3} = 5^{4-3} = 5$
5. $\frac{x^3}{x^7} = x^{3-7} = x^{-4} = \frac{1}{x^4}$
6. $7^{-2} = \frac{1}{7^2} = \frac{1}{49}$

PRACTICE 2

1. 4.3×10^{-3} d. 0.0043
2. 4.3×10^3 c. 4,300
3. 4.3×10^{-6} b. 0.0000043
4. 4.3×10^{-1} e. 0.43
5. 4.3×10^1 a. 43

PRACTICE 3

1. *Read and understand the question.* This question is looking for the product of two values that have the same base.

   Make a plan. Add the exponents to find the solution.

   Carry out the plan. The problem becomes $4^3 \times 4^5 = 4^{3+5} = 4^8$.

   Check your answer. To check your answer, divide the product by one of the factors that were multiplied in the question.

   $$\frac{4^8}{4^3} = 4^{8-3}$$
   $$= 4^5$$

   Because this was the other factor, this answer is checking.

2. *Read and understand the question.* This question is looking for the quotient of two values that have the same base.

   Make a plan. Subtract the exponents to find the solution.

   Carry out the plan. The problem becomes

   $$\frac{x^{10}}{x^3} = x^{10-3}$$
   $$= x^7$$

   Check your answer. To check your answer, multiply the quotient by the factor x^3.

   $$x^7 \times x^3 = x^{7+3}$$
   $$= x^{10}$$

   Because this was the first value given in the question, this answer is checking.

3. *Read and understand the question.* This question is looking for the product of two values that have the same base.

   Make a plan. Add the exponents to find the solution.

   Carry out the plan. The problem becomes

   $$y^{12} \times y^4 = y^{12+4}$$
   $$= y^{16}$$

   Check your answer. To check your answer, divide the product by one of the factors that were multiplied in the question.

   $$\frac{y^{16}}{y^{12}} = y^{16-12}$$
   $$= y^4$$

   Because this was the other factor, this answer is checking.

4. *Read and understand the question.* This question is looking for the quotient of two values that have the same base.

Make a plan. Subtract the exponents to find the solution.

Carry out the plan. The problem becomes

$$\frac{7^5}{7^8} = 7^{5-8}$$
$$= 7^{-3}$$
$$= \frac{1}{7^3}$$

Check your answer. To check your answer, multiply the quotient by the factor 7^8.

$$\frac{1}{7^3} \times 7^8 = \frac{7^8}{7^3}$$
$$= 7^{8-3}$$
$$= 7^5$$

Because this was the first value given in the question, this answer is checking.

5. *Read and understand the question.* This question is looking for the standard form when the number in scientific notation is given.

Make a plan. Use the value between one and 10 and move the decimal point the number of places given by the exponent. Use zeros as placeholders where necessary.

Carry out the plan. Take the decimal 4.23 and move the decimal point 6 places to the right. Add zeros as placeholders where necessary. The number becomes 4,230,000.

Check your answer. To check your answer, convert from standard form back to scientific notation. Place the decimal point between the 4 and the 2. Since the decimal point has moved six places to the left, the exponent of 10 will be 6. The scientific notation is 4.23×10^6, which was the value given in the question. This answer is checking.

6. *Read and understand the question.* This question is looking for the scientific notation when the standard form is given.

Make a plan. Form a number between one and 10, and then write this number multiplied by a power of 10. The exponent is the number of places the decimal point moves from the end of the number to between the 5 and the 6.

Carry out the plan. Write the decimal 5.6. Next, multiply this value by 10^{-4} because the decimal point moves 4 places to the right. The scientific notation is 5.6×10^{-4}.

Check your answer. To check your answer, put this number back into standard form. Take the decimal 5.6 and move the decimal 4 places to the left. Add zeros as placeholders where necessary. The number becomes 0.00056. This answer is checking.

applications of algebra

The human mind has never invented a
labor-saving machine equal to algebra.
—AUTHOR UNKNOWN

This lesson will cover some of the most common application questions involving algebra. Topics include consecutive integers, mixture problems, coin problems, age problems, and distance problems.

CONSECUTIVE INTEGER PROBLEMS

Consecutive integers are numbers in order one after the other. The integers 3, 4, and 5 are three consecutive integers. Because they are integers, there are no fractions or decimals.

Consecutive **odd** integers are numbers such as 1, 3, 5, and –11, –9, –7.
Consecutive **even** integers are numbers such as 4, 6, 8, and –20, –18, –16.

. . . .

TIP: Since consecutive integers are always one apart from each other, use the expressions *x*, *x* + 1, *x* + 2, and so on, to define the variables in a **consecutive integer** problem.

Since consecutive odd integers are always two apart from each other, and consecutive even integers are always two apart as well, use the expressions x, $x + 2$, $x + 4$, and so on, to define the variables in a **consecutive odd integer** or **consecutive even integer** problem.

Example 1

The sum of three consecutive integers is 60. What are the integers?

Read and understand the question. This question is looking for three consecutive integers that add to 60.

Make a plan. Use the expressions x, $x + 1$, and $x + 2$ to define the variables. Then, write an equation by adding these expressions and setting them equal to 60.

Carry out the plan. First, let x = the smallest integer, let $x + 1$ = the next integer, and let $x + 2$ = the greatest integer. Then, add these expressions together and set them equal to 60. The equation becomes $x + x + 1 + x + 2 = 60$. Combine like terms to get $3x + 3 = 60$. Subtract 3 from each side of the equation.

$$3x + 3 - 3 = 60 - 3$$

Simplify to get $3x = 57$

Divide each side by 3 to get the variable x alone.

$$\frac{3x}{3} = \frac{57}{3}$$
$$x = 19$$

Because $x = 19$, then $x + 1 = 20$ and $x + 2 = 21$. The three integers are 19, 20, and 21.

Check your answer. Check your answer by finding the sum of the three integers. The sum is $19 + 20 + 21 = 60$, so this answer is checking.

TIP: Be sure to use your equation solving skills as outlined in Lesson 13 to help with solving word problems with equations. Refer back to the details in that lesson when necessary.

Example 2

Twice the smaller of two consecutive even integers is equal to the larger even integer increased by 32. What are the two integers?

Read and understand the question. This question is looking for two consecutive even integers where clues are given for each.

Make a plan. Use the expressions x and $x + 2$ to define the variables. Then, write an equation by multiplying the smaller integer by 2 and setting this equal to the larger plus 32.

Carry out the plan. First, let x = the smaller even integer and let $x + 2$ = the larger even integer. Then, multiply the smaller by 2 to get $2x$ and increase the larger by 32 to get $(x + 2) + 32$. Set these values equal to each other. The equation becomes

$$2x = (x + 2) + 32$$

Combine like terms to get $2x = x + 34$. Subtract x from each side of the equation.

$$2x - x = x - x + 34$$

Simplify to get $x = 34$. Because $x = 34$, then $x + 2 = 36$. The two integers are 34 and 36.

Check your answer. Check your answer by making sure the two integers fit the clues in the problem. Twice the smaller is $2 \times 34 = 68$, and the larger increased by 32 is $36 + 32 = 68$. Because these values are equal, this answer is checking.

MIXTURE PROBLEMS

Mixture problems involve two or more different types of materials that will be combined to form one mixture. To solve mixture problems, use the clues in the problem to set up the *let* statements for the amounts of each type that will be mixed. Then, add the parts together and set the sum equal to the total mixture. Use the following example to help with these steps.

Example 3

Jane wants to mix $4 per pound coffee with $6 per pound coffee to get a mixture that costs $5.50 per pound. If the total mixture contains 10 pounds, how many pounds of the $4 per pound coffee should she buy?

Read and understand the question. Jane is mixing together two different types of coffee. Each type is a different price. You are looking for the number of pounds of the less expensive coffee as a final answer.

Make a plan. Write an expression for the amounts of each type of coffee using the information given. Then, multiply each by the price per pound. Set this equal to the total mixture, which is equal to $5.50 × 10 or 5.50(10).

Carry out the plan. Let x = the number of pounds of the $4 per pound coffee. Since there is a total of 10 pounds in all, then $10 - x$ = the number of pounds of the $6 per pound coffee. Next, write an equation that adds the cost of the two types of coffee and sets it equal to the total. The cost of the $4 coffee is $4x$, the cost of the $6 coffee is $6(10 - x)$, and the total cost is 5.50(10). The equation is

$$4x + 6(10 - x) = 5.50(10)$$

Simplify by using the distributive property:

$$4x + 60 - 6x = 55$$

Combine like terms:

$$-2x + 60 = 55$$

Subtract 60 from each side of the equation:

$$-2x + 60 - 60 = 55 - 60$$

Simplify to get $-2x = -5$. Divide each side by -2.

$$\frac{-2x}{-2} = \frac{-5}{-2}$$
$$x = 2.5$$

Because x represents the number of pounds of the $4 per pound coffee, Jane should buy 2.5 pounds.

Check your answer. To check this solution, first find the amount of the $6 per pound coffee. Using your solution, this should be 10 – 2.5 = 7.5 pounds. Next, multiply the number of pounds of each by the cost of each. $4 × 2.5 pounds is equal to $10 and $6 × 7.5 pounds is equal to $45. These amounts have a sum of $10 + $45 = $55, which was the cost of the total mixture. This answer is checking.

COIN PROBLEMS

Coin problems are similar to mixture problems in that you are taking different values of the various coins and *mixing* them together to get a total amount. Take the following example.

Sebastian has a total of $2.75 in his bank. He has only quarters and dimes, and has 10 more dimes than quarters. How many of each coin does he have in his bank?

Read and understand the question. This question is looking for the total number of quarters and the total number of dimes in a bank containing only quarters and dimes. There are 10 more dimes than quarters.

Make a plan. Write an expression for the number of quarters and the number of dimes in the bank. Multiply each of these expressions by the value of each type of coin. Then, add these expressions and set them equal to the total of $2.75.

Carry out the plan. Let x = the number of quarters and let $x + 10$ = the number of dimes. A quarter has a value of $0.25, so multiply 0.25 by x to get $0.25x$. A dime has a value of $0.10 so multiply 0.10 by $x + 10$ to get $0.10(x + 10)$. Next, write an equation by adding these expressions and setting them equal to $2.75. The equation is $0.25x + 0.10(x + 10) = 2.75$. Use the distributive property to eliminate the parentheses. The equation becomes $0.25x + 0.10x + 1 = 2.75$. Combine like terms to get $0.35x + 1 = 2.75$. Subtract 1 from each side to get $0.35x = 1.75$. Divide each side by 0.35.

$$\frac{0.35x}{0.35} = \frac{1.75}{0.35}$$
$$x = 5$$

Therefore, there are 5 quarters and 5 + 10 = 15 dimes.

Check your answer. Check this solution by making sure that the coins are equal to a total of $2.75. Five quarters has a value of $0.25 × 5 = $1.25, and 15 dimes has a value of $0.10 × 15 = $1.50. The sum is $1.25 + $1.50 = $2.75, so this answer is checking.

AGE PROBLEMS

To solve age problems, use the clues within the problem to define a variable for each person in the problem. Then, write an equation based on the given information about their ages.

Tasha is 6 years older than Frank. If the sum of their ages is 54, how old is Tasha?

Read and understand the question. This question is looking for Tasha's age when clues are given about her age related to Frank's age.

Make a plan. Write an expression for each person's age, and add these expressions together for a total of 54.

Carry out the plan. Let x = Frank's age. Because Tasha is 6 years older, let $x + 6$ = Tasha's age. Add the ages together and set the sum equal to 54, since the sum of their ages is 54.

$$x + x + 6 = 54$$

Combine like terms to get $2x + 6 = 54$. Subtract 6 from each side to get

$$2x + 6 - 6 = 54 - 6 = 48$$
$$2x = 48$$

Divide each side of the equation by 2 to get $x = 24$. Therefore, Frank is 24 years old and Tasha is $24 + 6 = 30$ years old.

Check your answer. To check this answer, add the ages and make sure they have a sum of 54 and that their difference is 6.

$$24 + 30 = 54$$

and

$$30 - 24 = 6$$

so this problem is checking.

D = R × T PROBLEMS (DISTANCE = RATE × TIME)

The formula *distance* = *rate* × *time*, or $d = r \times t$, is commonly used in both math and science. This formula states that the rate, or the speed, at which something travels multiplied by the time spent traveling is equal to the distance traveled. Use this formula to solve the following problem.

Two people leave from the same city and drive in opposite directions. Person A is traveling at a rate of 55 miles per hour and person B is traveling at a rate of 60 miles per hour. If they leave the city at the exact same time, how long will it take for them to be 345 miles apart?

Read and understand the question. This question is looking for the time it will take two people leaving from the same point in opposite directions to be 345 miles apart. Each person is driving at a different rate.

Make a plan. Use the formula *distance* = *rate* × *time*, or $d = r \times t$ to solve this problem. Write an expression for the distance of each person, showing that the sum of these distances is 345 miles.

Carry out the plan. Each person's rate is given and the time is unknown, so use t for the time. Since *distance* = *rate* × *time*, person A's distance is $55t$ and person B's distance is $60t$. Write an equation that adds these two distances and sets the sum equal to 345 miles.

$$55t + 60t = 345$$

Combine like terms to get $115t = 345$. Divide each side of the equation by 115.

$$\frac{115t}{115} = \frac{345}{115}$$
$$t = 3$$

In three hours, they will be 345 miles apart.

Check your answer. To check this problem, substitute $t = 3$ for the time, and make sure that the total distance between them adds to 345 miles.

$$55 \text{ miles per hour} \times 3 \text{ hours} = 165 \text{ miles}$$

and

$$60 \text{ miles per hour} \times 3 \text{ hours} = 180 \text{ miles}$$

This is a sum of $165 + 180 = 345$ miles, so this solution is checking.

PRACTICE

1. The sum of three consecutive odd integers is 135. What are the integers?

2. Three times a smaller of two consecutive integers is equal to twice the larger integer increased by 27. What are the integers?

3. Carolyn is buying two different amounts of candy to form a 6-pound mixture of candy that costs $3 per pound. If one type of candy costs $2.50 per pound, and the other type costs $4 per pound, how much of each type does she need to buy?

4. In her change purse, Josie has nickels and dimes. If she has a total of $2.40 and has twice as many nickels as dimes, how many of each coin does she have?

5. James is 10 years less than twice as old as Brandon. If the sum of their ages is 56, what are their ages?

6. Kaitlin's mother drives her to her grandparent's house 160 miles away and averages 64 miles per hour. On the return trip home, they average only 40 miles per hour because of traffic. How long did the return trip take?

ANSWERS

Practice

1. *Read and understand the question.* This question is looking for three consecutive odd integers that add to 135.
   Make a plan. Use the expressions x, $x + 2$, and $x + 4$ to define the variables. Then, write an equation by adding these expressions and setting them equal to 135.
   Carry out the plan. First, let x = the smallest odd integer, let $x + 2$ = the next odd integer, and let $x + 4$ = the greatest odd integer. Then, add these expressions together and set them equal to 135. The equation becomes $x + x + 2 + x + 4 = 135$. Combine like terms to get $3x + 6 = 135$. Subtract 6 from each side of the equation.
   $$3x + 6 - 6 = 135 - 6$$

Simplify to get $3x = 129$. Divide each side by 3 to get the variable x alone.

$$\frac{3x}{3} = \frac{129}{3}$$
$$x = 43$$

Because $x = 43$, then $x + 2 = 45$, and $x + 4 = 47$. The three integers are 43, 45, and 47.

Check your answer. Check your answer by finding the sum of the three odd integers. The sum is $43 + 45 + 47 = 135$, so this problem is checking.

2. *Read and understand the question.* This question is looking for two consecutive integers where clues are given for each.

   Make a plan. Use the expressions x and $x + 1$ to define the variables. Then, write an equation by multiplying the smaller integer by 3 and setting this equal to two times the larger plus 27.

   Carry out the plan. First, let $x =$ the smaller integer and let $x + 1 =$ the larger integer. Then, multiply the smaller by 3 to get $3x$ and increase twice the larger by 27 to get $2(x + 1) + 27$. Set these values equal to each other. The equation becomes $3x = 2(x + 1) + 27$. Apply the distributive property to get $3x = 2x + 2 + 27$. Simplify to get $3x = 2x + 29$. Subtract $2x$ from each side of the equation:

   $$3x - 2x = 2x - 2x + 29$$

   Simplify to get $x = 29$. Because $x = 29$, then $x + 1 = 30$. The two integers are 29 and 30.

   Check your answer. Check your answer by making sure the two integers fit the clues in the problem. Three times the smaller is $3 \times 29 = 87$, and twice the larger increased by 27 is $2(30) + 27 = 60 + 27 = 87$. Because these values are equal, this problem is checking.

3. *Read and understand the question.* Carolyn is mixing together two different types of candy. Each type is a different price. You are looking for the number of pounds of each as a final answer.

   Make a plan. Write an expression for the amounts of each type of candy using the information given, and multiply each by the price per pound. Set this equal to the total mixture, which is equal to 3.00×6 or $3(6)$.

   Carry out the plan. Let $x =$ the number of pounds of the $2.50 per pound candy. Since there is a total of 6 pounds in all, then $6 - x =$ the number of pounds of the $4 per pound candy. Next, write an equation that adds the cost of the two types of candy and sets it equal to the total. The cost of the $2.50 candy is $2.5x$, the cost of the $4 candy is $4(6 - x)$, and the total cost is $3(6)$. The equation is $2.5x + 4(6 - x) = 3(6)$. Simplify by using the distributive property:

   $$2.5x + 24 - 4x = 18$$

Combine like terms.

$$-1.5x + 24 = 18$$

Subtract 24 from each side of the equation.

$$-1.5x + 24 - 24 = 18 - 24$$

Simplify to get $-1.5x = -6$. Divide each side by -1.5:

$$\frac{-1.5x}{-1.5} = \frac{-6}{-1.5}$$

$$x = 4$$

Carolyn should buy 4 pounds of the $2.50 candy and $6 - 4 = 2$ pounds of the $4 per pound candy.

Check your answer. To check this solution, multiply the number of pounds of each by the cost of each and see if it is a total of $\$3 \times 6 = \$18$.

$\$2.5 \times 4$ pounds is equal to $10 and $\$4 \times 2$ pounds is equal to $8. These amounts have a sum of $\$10 + \$8 = \$18$, which was the cost of the total mixture. This answer is checking.

4. *Read and understand the question.* This question is looking for the total number of nickels and the total number of dimes in a coin purse containing only nickels and dimes. There are twice as many nickels as dimes.

Make a plan. Write an expression for the number of dimes and the number of nickels in the bank. Multiply each of these expressions by the value of each type of coin. Then, add these expressions and set them equal to the total of $2.40.

Carry out the plan. Let x = the number of dimes, and let $2x$ = the number of nickels. A dime has a value of $0.10, so multiply 0.10 by x to get $0.10x$. A nickel has a value of $0.05 so multiply 0.05 by $2x$ to get $0.05(2x)$. Next, write an equation by adding these expressions and setting them equal to $2.40. The equation is $0.10x + 0.05(2x) = 2.40$. Multiply to eliminate the parentheses. The equation becomes $0.10x + 0.10x = 2.40$. Combine like terms to get $0.20x = 2.40$. Divide each side by 0.20:

$$\frac{0.20x}{0.20} = \frac{2.40}{0.20}$$

$$x = 12$$

Therefore, there are 12 dimes and $2(12) = 24$ nickels.

Check your answer. Check this solution by making sure that the coins equal a total of $2.40. Twenty-four nickels has a value of $\$0.05 \times 24 = \$1.20$, and 12 dimes has a value of $\$0.10 \times 12 = \$1.20$. The sum is $\$1.20 + \$1.20 = \$2.40$, so this answer is checking.

5. *Read and understand the question.* This question is looking for both ages when clues are given about how their ages are related. The sum total of their ages is also given.

Make a plan. Write an expression for each person's age, and add these expressions together for a total of 56.

Carry out the plan. Let x = Brandon's age. Because James's age is 10 years less than twice Brandon's age, let $2x - 10$ = James's age. Add the ages together and set the sum equal to 56, since the sum of their ages is 56:

$$x + 2x - 10 = 56$$

Combine like terms to get $3x - 10 = 56$. Add 10 to each side to get

$$3x - 10 + 10 = 56 + 10$$
$$3x = 66$$

Divide each side of the equation by 3 to get $x = 22$. Therefore, Brandon is 22 years old and James is $2(22) - 10 = 44 - 10 = 34$ years old.

Check your answer. To check this answer, add the ages and make sure they have a sum of 56: $22 + 34 = 56$, so this answer is checking.

6. *Read and understand the question.* This question is looking for the time it took on the return trip. Be aware of extra information. The fact that they averaged 64 miles per hour on the way to Kaitlin's grandmother's house is not necessary to find the solution to the problem.

Make a plan. Use the formula *distance = rate × time*, or $d = r \times t$ to solve this problem. Write an expression using the distance of the return trip and the rate to find the time.

Carry out the plan. The time is unknown, so use t for the time. Since *distance = rate × time*, the return trip of 160 miles at a rate of 40 miles per hour gives the equation $40t = 160$. Divide each side of the equation by 40.

$$\frac{40t}{40} = \frac{160}{40}$$
$$t = 4$$

It took 4 hours for the return trip.

Check your answer. To check this problem, substitute $t = 4$ for the time and make sure that the total distance is 160 miles:

$$40 \text{ miles per hour} \times 4 \text{ hours} = 160 \text{ miles}$$

so this solution is checking.

S E C T I O N 4

geometry and measurement word problems— measuring up

GEOMETRY IS THE STUDY of one- and two-dimensional figures in the world around us. Just take a look, and geometric shapes are everywhere! The real world is comprised of all the shapes and properties of geometry, and it is important to know when and how to apply these properties. Although it is extremely useful, applying the concepts of geometry is not always simple. There are numerous properties and formulas to know that are used in geometry. In the same way, word problems involving geometry can be challenging, especially when a diagram is not given. Be sure to use the strategies of drawing a picture, writing an equation, using a formula, or any other problem solving strategy, if it makes the problem easier to solve.

The word problems and strategies in this section are focused on geometry and measurement and will use these important properties in many different real-world applications.

This section will explain geometry and measurement word problems including:

- types of angles
- triangles
- quadrilaterals
- similar figures
- Pythagorean theorem

- perimeter
- circumference
- area
- surface area
- volume
- coordinate geometry

angle word problems

The composer opens the cage door for arithmetic,
the draftsman gives geometry its freedom.
—Cocteau (1839–1963)

This lesson will review the basic terms of geometry and list the special types of angles and angle pairs found in many geometry word problems.

BASIC FIGURES OF GEOMETRY

Knowing the basic terms of geometry can make the study of more complicated shapes much easier. Listed next are a few of these basic terms.

A **point** is a location in space.

A **line** is an infinite collection of points extending in opposite directions.

A **plane** is a never-ending flat surface that extends in all directions.

A **ray** is the set of all points extending in a straight line from one side of an endpoint.

ANGLES

An **angle** is made up two rays that meet at a common endpoint. The rays are the sides of the angle, and the common endpoint is the **vertex** of the angle.

Angles can be named in various ways. They can be named by the letter of the vertex, by three letters with the middle letter being the vertex, or by a number written in the interior of the angle. These three ways to name angles ($\angle$) are shown in the following figure.

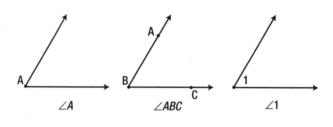

$\angle A$ $\angle ABC$ $\angle 1$

· ·

TIP: Here are some important types of angles and their measures:

Acute angles measure less than 90°.
Right angles measure 90°.
Obtuse angles measure greater than 90 but less than 180°.
Straight angles measure 180°.
Reflex angles measure more than 180°.

· ·

There are special types of pairs of angles that are common to geometry word problems. These special types are explained next.

Adjacent Angles

Adjacent angles are angles next to each other that share a common ray, or side, and a common vertex. Nonadjacent angles are not next to each other and do not share a common ray, or side. An example of each is shown below.

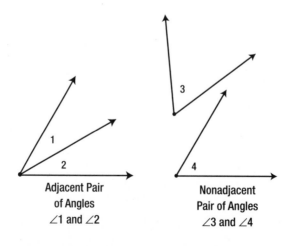

Adjacent Pair
of Angles
$\angle 1$ and $\angle 2$

Nonadjacent
Pair of Angles
$\angle 3$ and $\angle 4$

Vertical Angles

Vertical angles are the nonadjacent angles formed by two intersecting lines. The measures of vertical angles are always equal.

Look at the following example. The measure of angles 1 and 3 are each 60°; they are vertical angles. The measure of angles 2 and 4 are each 120°; they are also a vertical pair of angles. Notice that the adjacent angles have a sum of $120 + 60 = 180°$.

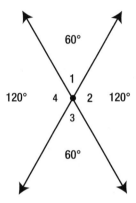

Complementary and Supplementary Pairs of Angles

Complementary angles are two angles that have a sum of 90°. They can be adjacent, or nonadjacent angles, as shown next.

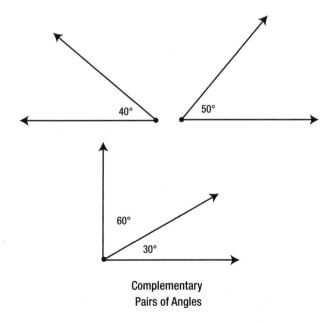

Complementary
Pairs of Angles

Supplementary angles are two angles that have a sum of 180°. They can be adjacent, or nonadjacent angles, as shown next. If two supplementary angles are also adjacent, they are known as a **linear pair**.

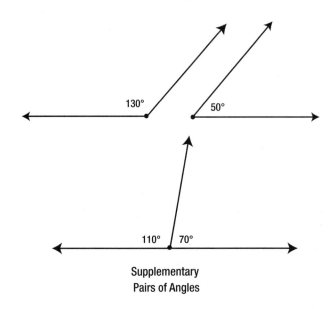

Supplementary
Pairs of Angles

PRACTICE 1

1. The measure of an angle is 31°. This angle is classified as an _____ angle.

2. The sum of the measures of two angles is exactly 90°. These are known as a pair of _____ angles.

3. One angle with a measure of 90° is known as a _____ angle.

4. An angle with a degree measure between 90° and 180° is an _____ angle.

5. If one angle of a linear pair is 30°, then the measure of the other angle in the pair is _____ °.

ANGLES FORMED BY TWO PARALLEL LINES CUT BY A TRANSVERSAL

When two parallel lines are cut by a transversal, there are many patterns in the angles, as shown next.

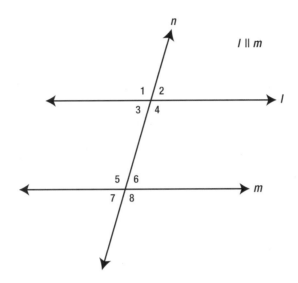

Vertical Angles: As explained in the previous section, angle pairs 1 and 4, 2 and 3, 5 and 8, and 6 and 7 are vertical angles.

Alternate Interior Angles: These are the angles on the interior of the parallel lines, but on opposite sides of the transversal. These are the angle pairs 4 and 5 and 3 and 6.

Alternate Exterior Angles: These are the angles on the exterior of the parallel lines, but on opposite sides of the transversal. These are the angle pairs 1 and 8 and 2 and 7.

Corresponding Angles: These are the nonadjacent angles on the same side of the transversal, but one is an interior angle and the other is an exterior angle. Corresponding angles are named by the pairs 1 and 5, 2 and 6, 3 and 7, and 4 and 8.

Supplementary Angles: Any two adjacent angles in the diagram are supplementary. Some of these pairs are 1 and 2, 2 and 4, 3 and 4, 1 and 3, 5 and 6, and so on.

TIP: When dealing with parallel lines cut by a transversal, the following angle pairs have equal measure.

Corresponding Angles
Vertical Angles
Alternate Interior Angles
Alternate Exterior Angles

Adjacent angles are **supplementary**; they add to 180°.

PRACTICE 2

Using the following figure, find the measure of each angle.

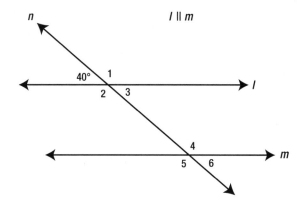

$m\angle 1 =$ _____

$m\angle 2 =$ _____

$m\angle 3 =$ _____

$m\angle 4 =$ _____

$m\angle 5 =$ _____

$m\angle 6 =$ _____

WORD PROBLEMS WITH SPECIAL ANGLE PAIRS

Vertical Angles

Vertical angle problems can be solved by setting the values for each angle equal to each other.

Example
A pair of vertical angles are represented by the expressions $x + 30$ and $5x - 10$. What is the measure in degrees of each angle?

Read and understand the question. This question is looking for the measure of each vertical angle. The vertical angles formed by two intersecting lines are always congruent.

Make a plan. Set the expressions that represent each angle equal to each other, and solve for x.

Carry out the plan. The equation is $x + 30 = 5x - 10$. Subtract x from each side to get $30 = 4x - 10$. Then, add 10 to each side to simplify the equation to $40 = 4x$. Divide each side by 4 to get the variable alone: $x = 10$. Therefore, the angles are $10 + 30 = 40°$ each.

Check your answer. To check this solution, substitute $x = 10$ into the other expression to be sure it also is equal to $40°$: $5(10) - 10 = 50 - 10 = 40°$. Each vertical angle is $40°$, so this answer is checking.

Complementary Angles

Complementary angle problems can be solved by adding the values of the angles and setting the sum equal to $90°$.

Example
One angle of a complementary pair is equal to twice the measure of the other angle. What is the measure in degrees of each angle?

Read and understand the question. This question is looking for the measure of each complementary angle. The measure of two complementary angles is always $90°$.

Make a plan. Write the expression for each angle and set the sum equal to 90.

Carry out the plan. Let x = the smaller angle and let $2x$ = the larger angle. Therefore, the equation is $x + 2x = 90$. Combine like terms to get $3x = 90$. Divide each side of the equation by 3 to get the variable alone: $x = 30$. Thus, $2x = 60$. The two angles measure 30° and 60°, respectively.

Check your answer. To check this problem, make sure that the sum of the measures of the angles is equal to 90° and that one angle is twice the other. The angles are $30 + 60 = 90°$ and $(30)(2) = 60$, so this answer is checking.

Supplementary Angles

Supplementary angle problems can be solved by adding the values of the angles and setting the sum equal to 180°.

Example
One of two supplementary angles measures 50° more than the other measures. What is the measure in degrees of each angle?

Read and understand the question. This question is looking for the measure of each supplementary angle. The measure of two supplementary angles is always 180°.

Make a plan. Write the expression for each angle and set the sum equal to 180.

Carry out the plan. Let x = the smaller angle and let $x + 50$ = the larger angle. Therefore, the equation is $x + x + 50 = 180$. Combine like terms to get $2x + 50 = 180$. Subtract 50 from each side of the equation to simplify it to $2x = 130$. Divide each side of the equation by 2 to get the variable alone: $x = 65$. Thus, $x + 50 = 115$. The two angles measure 65° and 115°, respectively.

Check your answer. To check this problem, make sure that the sum of the measures of the angles is equal to 180° and that one angle is 50° more than the other. The angles are $65 + 115 = 180°$, and $115 - 65 = 50$, so this answer is checking.

Alternate Interior and Alternate Exterior Angles

Alternate interior and alternate exterior angle problems can be solved by setting the values for each angle equal to each other.

Example

When parallel lines are cut by a transversal, the measure of one alternate interior angle is equal to 60 less than 3 times the other. Find the measure of both angles.

Read and understand the question. This question is looking for the measure of each alternate interior angle. The alternate interior angles formed by two parallel lines cut by a transversal are congruent.

Make a plan. Write an expression to represent each angle. Then, set the expressions equal to each other and solve for x.

Carry out the plan. Let x = one angle and $3x - 60$ = the other angle. The equation is $x = 3x - 60$. Add 60 to each side of the equation to get $x + 60 = 3x$. Subtract x from each side to get $60 = 2x$. Then, divide each side of the equation by 2 to get the variable alone: $x = 30$. Therefore, the angles are 30° each.

Check your answer. To check this solution, substitute $x = 30$ into the other expression to be sure it also is equal to 30°: $3(30) - 60 = 90 - 60 = 30°$. Each alternate interior angle is 30°, so this answer is checking.

Corresponding Angles

Corresponding angle problems can be solved by setting the values for each angle equal to each other.

Example

Two parallel lines are cut by a transversal. The sum of two corresponding angles formed is 110°. What is the measure in degrees of each angle?

Read and understand the question. This question is looking for the measure of each corresponding angle. When parallel lines are cut by a transversal, the measures of corresponding angles are equal.

Make a plan. Take the given sum and divide by 2.

Carry out the plan. 110° divided by 2 is 55°. Each corresponding angle is 55°.

Check your answer. To check this solution, make sure that the sum of the two angles is 110°: 55 + 55 = 110, so this answer is checking.

PRACTICE 3

1. The measures of two vertical angles are expressed as $6x$ and $2x + 80$. What is the measure in degrees of each angle?

2. The measure of an angle is 15 more than its complement. What is the measure in degrees of both angles?

3. The measure of an angle is twice the measure of its supplement. What is the measure in degrees of both angles?

4. The measure of two alternate exterior angles formed by two parallel lines cut by a transversal are expressed as $5x + 10$ and $2x + 55$. What is the number of degrees in each angle?

5. Two parallel lines are cut by a transversal. A pair of corresponding angles is represented by the expressions $7x$ and $5x + 40$. What is the measure of each angle?

ANSWERS

Practice 1

1. acute
2. complementary
3. right
4. obtuse
5. $180 - 30 = 150$

Practice 2

$m\angle 1 = $ __ 140 __

$m\angle 2 = $ __ 140 __

$m\angle 3 = $ __ 40 __

$m\angle 4 = $ __ 140 __

$m\angle 5 = $ __ 140 __

$m\angle 6 = $ __ 40 __

Practice 3

1. *Read and understand the question.* This question is looking for the measure of each vertical angle. The vertical angles formed by two intersecting lines are always congruent.

   Make a plan. Use the expressions given that represent each angle, set the expressions equal to each other, and solve for x.

   Carry out the plan. Let $6x = $ one angle and $2x + 80 = $ the other angle. The equation is $6x = 2x + 80$. Subtract $2x$ from each side to get $4x = 80$. Then, divide each side by 4 to get the variable alone: $x = 20$. Therefore, the angles are $6(20) = 120°$ each.

   Check your answer. To check this solution, substitute $x = 20$ into the other expression to be sure it also is equal to $120°$: $2(20) + 80 = 40 + 80 = 120°$. Each vertical angle is $120°$, so this answer is checking.

2. *Read and understand the question.* This question is looking for the measure of each complementary angle. The sum of the measures of two complementary angles is always $90°$.

   Make a plan. Write the expression for each angle and set the sum equal to 90.

   Carry out the plan. Let $x = $ the smaller angle, and let $x + 15 = $ the larger angle. Therefore, the equation is $x + x + 15 = 90$. Combine like terms to get $2x + 15 = 90$. Subtract 15 from each side of the equation to get $2x = 75$. Divide each side of the equation by 2 to get the variable alone: $x = 37.5$. Thus, $x + 15 = 52.5$. The two angles measure $37.5°$ and $52.5°$, respectively.

   Check your answer. To check this problem, make sure that the sum of the measures of the angles is equal to $90°$ and that one angle is 15 more than the other. The angles are $37.5 + 52.5 = 90°$ and $52.5 - 37.5 = 15$, so this problem is checking.

3. *Read and understand the question.* This question is looking for the measure of each supplementary angle. The sum of the measures of two supplementary angles is always 180°.

Make a plan. Write the expression for each angle and set the sum equal to 180.

Carry out the plan. Let x = the smaller angle, and let $2x$ = the larger angle. Therefore, the equation is $x + 2x = 180$. Combine like terms to get $3x = 180$. Divide each side of the equation by 3 to get the variable alone: $x = 60$. Thus, $2x = 120$. The two angles measure 60° and 120°, respectively.

Check your answer. To check this problem, make sure that the sum of the measures of the angles is equal to 180° and that one angle is twice the other. The angles are $60 + 120 = 180°$ and $(60)(2) = 120$, so this answer is checking.

4. *Read and understand the question.* This question is looking for the measure of each alternate exterior angle. The alternate exterior angles formed by two parallel lines cut by a transversal are congruent.

Make a plan. Use the expressions that represent each angle. Set the expressions equal to each other and solve for x.

Carry out the plan. Let $5x + 10$ = one angle and let $2x + 55$ = the other angle. The equation is $5x + 10 = 2x + 55$. Subtract $2x$ from each side to get $3x + 10 = 55$. Subtract 10 from each side to get the equation $3x = 45$. Then, divide each side of the equation by 3 to get the variable alone: $x = 15$. Therefore, the angles are $5(15) + 10 = 75 + 10 = 85°$ each.

Check your answer. To check this solution, substitute $x = 15$ into the other expression to be sure it also is equal to 85°: $2(15) + 55 = 30 + 55 = 85°$. Each alternate exterior angle is 85°, so this solution is checking.

5. *Read and understand the question.* This question is looking for the measure of each corresponding angle. The corresponding angles formed by two parallel lines cut by a transversal are congruent.

Make a plan. Use the expressions that represent each angle. Set the expressions equal to each other, and solve for x.

Carry out the plan. Let $7x$ = one angle, and let $5x + 40$ = the other angle. The equation is $7x = 5x + 40$. Subtract $5x$ from each side to get $2x = 40$. Divide each side of the equation by 2 to get the variable alone: $x = 20$. Therefore, the angles are $7(20) = 140°$ each.

Check your answer. To check this solution, substitute $x = 20$ into the other expression to be sure it also is equal to 140°: $5(20) + 40 = 100 + 40 = 140°$. Each corresponding angle is 140°, so this solution is checking.

triangle word problems

[The universe] cannot be read until we have learnt the language
and become familiar with the characters in which it is written.
It is written in mathematical language, and the letters are
triangles, circles and other geometrical figures, without which
means it is humanly impossible to comprehend a single word.
—GALILEO GALILEI (1564–1642)

This lesson provides a review of the classification of triangles by their sides and angles. Triangle word problems will be explained, and you will practice solving them.

CLASSIFYING TRIANGLES BY THEIR ANGLES

Triangles can be classified, or named, based on their interior angles. The sum of the interior angles of any triangle is always 180°.

Acute Triangles

An **acute triangle** is a triangle where each angle measures less than 90°.
 The following triangle is an acute triangle.

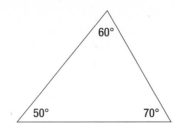

Example

The measures of two angles of an acute triangle are 55 and 60, respectively. What is the measure of the third angle?

Read and understand the question. This question is looking for the measure of the third angle of a triangle when the measures of two angles are given.

Make a plan. The triangle is acute, so the result will be an angle with measure less than 90°. Add the measures of the two known angles and subtract the sum from 180°.

Carry out the plan. The sum of the two known angles is 55 + 60 = 115: 180 – 115 = 65. The measure of the third angle is 65°.

Check your answer. To check this solution, add the three angles and make sure that the sum is 180°: 55 + 60 + 65 = 180. This solution is checking.

Obtuse Triangles

An **obtuse triangle** is a triangle where one angle measures more than 90° but less than 180°.

The following triangle is an example of an obtuse triangle.

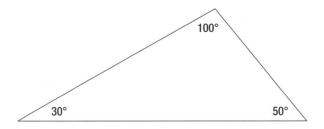

Example

The measures of two angles of an obtuse triangle are 110 and 30, respectively. What is the measure of the third angle?

Read and understand the question. This question is looking for the measure of the third angle of an obtuse triangle when two angles are given.

Make a plan. Add the two known angles and subtract this sum from 180°.

Carry out the plan. First, add the two known angles: 110 + 30 = 140. Then, subtract this amount from the total of 180°: 180 – 140 = 40. The third angle is 40°.

Check your answer. To check this solution, add the three angles to make sure that the sum is exactly 180°: 110 + 30 + 40 = 180. This answer is checking.

Right Triangles

A **right triangle** is a triangle where one angle measures 90°. The sides of the triangle that form the right angle are called the **legs**, and the side opposite the right angle is called the **hypotenuse**. In a right triangle, the two angles other than the right angle will each be acute.

The following triangle is a right triangle.

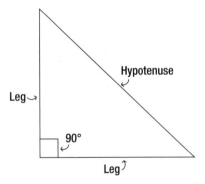

Example
The measure of one angle of a right angle is 30°. What is the measure of the other angles?

Read and understand the question. This question is asking for the measure of the third angle of a right triangle when one of the other acute angles is known.

Make a plan. Add the two known angles and subtract this sum from 180°.

Carry out the plan. Because the triangle is a right triangle, one of the other angles is 90°. Add the two known angles: 30 + 90 = 120. Finally, subtract this amount from the total of 180°: 180 – 120 = 60. The third angle is 60°.

Check your answer. To check this solution, add the three angles to make sure that the sum is exactly 180°: 30 + 90 + 60 = 180°. This solution is checking.

PRACTICE 1

1. In an acute triangle, the measures of two angles are 50° and 60°. What is the measure of the third angle?

2. In an obtuse triangle, the measures of two angles are 120° and 10°. What is the measure of the third angle?

3. One acute angle of a right triangle measures 35°. What is the measure of the other acute angle?

CLASSIFYING TRIANGLES BY THEIR SIDES

Triangles can also be classified, or named, by the number of congruent sides that they have.

Equilateral Triangles

An **equilateral triangle** is a triangle where all three sides of the triangle are congruent, or equal in length. Each of the angles of an equilateral triangle is always 180 ÷ 3 = 60°. An example of an equilateral triangle follows.

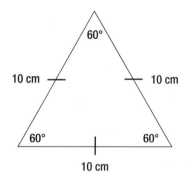

Example

The sides of an equilateral triangle have a sum of 60 cm. What is the measure of each side?

Read and understand the question. This question is asking for the measure of one side of an equilateral triangle. The sum of the three sides is given.

Make a plan. The triangle is equilateral, so each of the three sides has the same measure. Divide the sum of the three sides by 3 to find the length of one side.

Carry out the plan. If the sum of the measures of each side is 60 cm, divide 60 by 3 to find the length of one side: $60 \div 3 = 20$ cm. Each side is 20 cm.

Check your answer. To check this solution, multiply the length of one side by 3 to make sure the total is 60 cm: $20 \times 3 = 60$, so this answer is checking.

Isosceles Triangles

Isosceles triangles have two sides that are congruent. The two congruent sides meet the third side to form the congruent base angles. The third angle is known as the vertex angle. An example of an isosceles triangle is shown next.

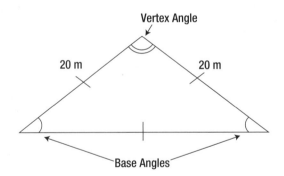

Example

The two congruent sides of an isosceles triangle are represented by the expressions $8x + 5$ and $9x - 7$. What is the value of x?

Read and understand the question. This question is looking for the value of x when expressions for two congruent sides of a triangle are given.

Make a plan. Set the expressions for the congruent sides equal to each other. Then, use the rules for equation solving to solve for x.

Carry out the plan. To find the value of x, set the expressions equal to each other and solve the equation. The equation is $8x + 5 = 9x - 7$. Subtract $8x$ from each side of the equation to get $5 = x - 7$. Add 7 to each side of the equation to get $x = 12$.

Check your answer. To check this solution, substitute $x = 12$ into each expression to make sure each one results in the same value.

$$8(12) + 5 = 96 + 5 = 101$$

$$9(12) - 7 = 108 - 7 = 101$$

This result is checking.

Scalene Triangles

A scalene triangle has sides that have three different lengths; in other words, there are no congruent sides. An example of a scalene triangle is shown in the following figure.

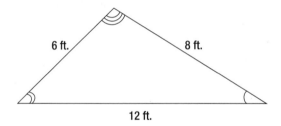

Example

In a scalene triangle, the ratio of the three sides is 3:4:5. If the sum of the sides is 24 units, what is the measure of each side?

Read and understand the question. This question is looking for the measure of each side of a scalene triangle when the ratio of the measures and the sum of the measures are given.

Make a plan. Because the triangle is scalene, each of the sides will have a different length. Use the strategies from ratios and equation solving to help with this question.

Carry out the plan. Multiply each of the values in the ratio by x to get $3x:4x:5x$. Next, set the sum of these terms equal to the sum of the sides, 24. The equation is

$$3x + 4x + 5x = 24$$

Combine like terms to get the equation $12x = 24$. Divide each side of the equation by 12 to get $x = 2$. Thus, the sides are $3(2) = 6$ units, $4(2) = 8$ units, and $5(2) = 10$ units.

Check your answer. To check this answer, add the three sides to be sure that the sum is 24: $6 + 8 + 10 = 24$, so this answer is checking.

....

TIP: Some triangles can be classified by **both** their **angles** and their **sides**. For example, an isosceles right triangle is a triangle with one right angle and two congruent sides. A scalene obtuse triangle is a triangle with one obtuse angle and three sides of different lengths.

....

TRIANGLE INEQUALITY

The **triangle inequality** is a property that states that the two shorter sides of any triangle must have a sum that is greater than the longest side of the triangle.

For example, the numbers 3, 4, and 5 could represent the sides of a triangle because the two shorter sides have a sum of $3 + 4 = 7$, which is greater than the longest side, 5.

The numbers 5, 7, and 13 could not represent the sides of a triangle because the two shorter sides have a sum of $5 + 7 = 12$, which is less than the longest side, 13.

....

TIP: When two sides of a triangle are known, the third side must be between the sum and the difference of these two known sides. In other words, subtract the two known values and then add the two known values. The third side must be between these two values.

....

Example
Two sides of a triangle are 6 units and 10 units. Between what two numbers must the third side lie?

Read and understand the question. This question is looking for the two numbers that the third side of a triangle must be between. Two sides of the triangle are known.

Make a plan. To find the values, add and subtract the given numbers.

Carry out the plan. To find this range, subtract the two known sides and add the two known sides. Because $10 - 6 = 4$, and $6 + 10 = 16$, the third side must be between 4 units and 16 units.

Check your answer. To check your solution, test a value in between 4 and 16. Take the number 9. If the third side measures 9 units, then the three sides are 6, 9, and 10. The sum of the two shorter sides is $6 + 9 = 15$, which is more than 10. This answer is checking.

PRACTICE 2

1. In a scalene triangle, one side is 5 units more than twice the measure of the shortest side. The other side is three times the measure of the shortest side. If the sum of the three sides is 41 units, what is the measure of each side of the triangle?

2. Two sides of an equilateral triangle measure $x + 1$ units and $2x - 10$ units, respectively. What is the value of x?

3. The two congruent sides of an isosceles triangle are represented by $2x - 5$ units and $x + 3$ units. What is the length of each of these congruent sides?

4. Two sides of a triangular garden measure 12 m and 15 m. Between what two values must the third side measure?

ANSWERS

Practice 1

1. *Read and understand the question.* This question is looking for the measure of the third angle of a triangle where the measures of two angles are given. *Make a plan.* The triangle is acute, so the result will be an angle with a measure less than 90°. Add the measures of the two known angles, and subtract the sum from 180°. *Carry out the plan.* The sum of the two known angles is

$$50 + 60 = 110$$
$$180 - 110 = 70$$

The measure of the third angle is 70°.

Check your answer. To check this solution, add the three angles and make sure that the sum is 180°.

$$50 + 60 + 70 = 180$$

This solution is checking.

2. *Read and understand the question.* This question is looking for the measure of the third angle of an obtuse triangle when two angles are given.
Make a plan. Add the two known angles and subtract this sum from 180°.
Carry out the plan. First, add the two known angles: $120 + 10 = 130$. Then, subtract this amount from the total of 180°: $180 - 130 = 50$. The third angle is 50°.
Check your answer. To check this solution, add the three angles to make sure that the sum is exactly 180°: $120 + 10 + 50 = 180°$. This answer is checking.

3. *Read and understand the question.* This question is asking for the measure of the third angle of a right triangle when one of the other acute angles is known.
Make a plan. Add the two known angles and subtract this sum from 180°.
Carry out the plan. Because the triangle is a right triangle, one of the other angles is 90°. Then, add the two known angles: $35 + 90 = 125$. Finally, subtract this amount from the total of 180°: $180 - 125 = 55$. The third angle is 55°.
Check your answer. To check this solution, add the three angles to make sure that the sum is exactly 180°: $35 + 90 + 55 = 180°$. This question is checking.

Practice 2

1. *Read and understand the question.* This question is looking for the measure of each side of a scalene triangle when clues about the measures and the sum of the measures are given.
Make a plan. Because the triangle is scalene, each of the sides will have a different length. Use the strategies from equation solving to help with this question.
Carry out the plan. Let x = the length of the shortest side. Since one side is five more than twice the shortest, $2x + 5$ = the length of the second side. The third side is three times the measure of the shortest side, so $3x$ = the

length of the third side. Next, set the sum of these expressions equal to the sum of the sides, 41. The equation is

$$x + 2x + 5 + 3x = 41$$

Combine like terms to get the equation $6x + 5 = 41$. Subtract 5 from each side of the equation to get $6x = 36$. Divide each side of the equation by 6 to get $x = 6$. Thus, the sides are $x = 6$ units, $2(6) + 5 = 12 + 5 = 17$ units, and $3(6) = 18$ units.

Check your answer. To check this answer, add the three sides to be sure that the sum is 41: $6 + 17 + 18 = 41$, so this answer is checking.

2. *Read and understand the question.* This question is asking for the value of x when two expressions of the measure of a side of an equilateral triangle are given.

Make a plan. The triangle is equilateral, so set the expressions for the congruent sides equal to each other. Then, use the rules for equation solving to solve for x.

Carry out the plan. To find the value of x, set the expressions equal to each other and solve the equation. The equation is $x + 1 = 2x - 10$. Subtract x from each side of the equation to get $1 = x - 10$. Add 10 to each side of the equation to get $11 = x$.

Check your answer. To check this solution, substitute $x = 11$ into each expression to make sure they result in the same value:

$$11 + 1 = 12$$
$$2(11) - 10 = 22 - 10 = 12$$

This result is checking.

3. *Read and understand the question.* This question is looking for the value of x when expressions for two congruent sides of a triangle are given.

Make a plan. Set the expressions for the congruent sides equal to each other. Then, use the rules for equation solving to solve for x.

Carry out the plan. To find the value of x, set the expressions equal to each other and solve the equation. The equation is $2x - 5 = x + 3$. Subtract x from each side of the equation to get $x - 5 = 3$. Add 5 to each side of the equation to get $x = 8$. Substitute this value into one of the expressions to find the length of a side:

$$2(8) - 5 = 16 - 5 = 11$$

Each side is 11 units in length.

Check your answer. To check this solution, substitute $x = 8$ into the other expression to make sure it results in the same value: $8 + 3 = 11$. This result is checking.

4. *Read and understand the question.* This question is looking for the two numbers that the third side of a triangle must be between. Two sides of the triangle are known.

Make a plan. To find the values, add and subtract the given numbers.

Carry out the plan. To find this range, subtract the two known sides and add the two known sides. Because $15 - 12 = 3$, and $15 + 12 = 27$, the third side must be between 3 m and 27 m.

Check your answer. To check your solution, test a value between 3 and 27. Take the number 10. If the third side measures 10 units, then the three sides are 10, 12, and 15. The sum of the two shorter sides is $10 + 12 = 22$, which is more than 15. This answer is checking.

quadrilateral word problems

[G]eometry is not true, it is advantageous.
—HENRI POINCARÉ (1854–1912)

This lesson will review the properties of quadrilaterals and the special types of quadrilaterals. Word problems involving quadrilaterals will be solved.

QUADRILATERALS

Quadrilaterals are four-sided polygons. This means that they are closed figures with four line segments as sides. The interior angles of every quadrilateral total 360°.

There are many types of special quadrilaterals. These quadrilaterals are classified, or named, based on their properties.

The first group of special quadrilaterals is the parallelograms.

Parallelogram

A **parallelogram** is a quadrilateral with opposite sides congruent, or the same measure. Parallelograms also have opposite angles congruent, and the diagonals bisect each other. In addition, consecutive angles of any parallelogram are supplementary. In other words, angles that are next to each other, like angle *A* and angle *B* in the following figure, add to 180 degrees. This property and others are shown in the examples below.

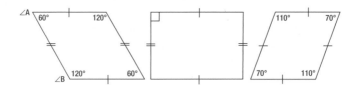

Rhombus

A **rhombus** is a parallelogram with all four sides congruent. It could look like a square that is leaning over. Keep in mind that all squares are also rhombuses. Rhombuses have all the properties of parallelograms, in addition to the fact that the diagonals are perpendicular, or meet at right angles. Examples of rhombuses are shown in the following figures.

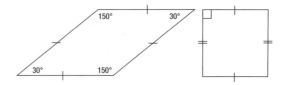

Rectangle

A **rectangle** is a parallelogram with four right angles. It could look like a parallelogram standing up straight. Rectangles also have all of the properties of parallelograms, in addition to the fact that the diagonals are congruent. Following are two examples of rectangles.

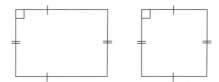

Square

A **square** is a rhombus with right angles. Therefore, all sides are congruent, and all angles are right angles. Squares have all of the properties of rhombuses, in addition to having four congruent angles and congruent diagonals. A square is shown next.

··

TIP: A **square** can be remembered as a rectangle with four congruent sides, or a rhombus with four right angles.

··

The following practice questions test your knowledge of the properties of quadrilaterals in general and parallelograms.

PRACTICE 1

1. One side of a square is 4 meters. What is the measure of each of the other sides?

2. What is the value of x in the following figure?

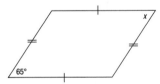

3. The measure of a side of a parallelogram is 24 inches. What is the measure of the side opposite from this side?

4. The measure of one angle of a rhombus is 35°. What is the measure of a consecutive angle of that angle?

The second group of special quadrilaterals is the trapezoids.

TRAPEZOID

A **trapezoid** is a quadrilateral with exactly one pair of parallel sides. Trapezoids have two parallel sides that are not the same measure; these sides are called the bases of the trapezoid. The two sides that are not parallel are called the legs of the trapezoid. Trapezoid examples are shown in the following figures.

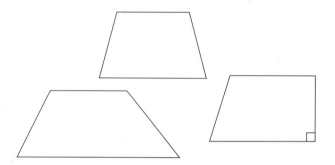

Isosceles Trapezoid

An **isosceles trapezoid**, like an isosceles triangle, has two sides congruent. In this type of trapezoid, the legs are the same measure. Isosceles triangles also have congruent base angles and congruent diagonals. These properties are shown in the following figure.

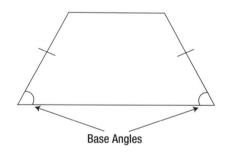

Base Angles

....

TIP: All **parallelograms** have two pairs of parallel sides.
All **trapezoids** have only one pair of parallel sides.

....

This practice set will test your skills applying the properties of trapezoids.

PRACTICE 2

1. Three angles of a trapezoid measure 100°, 90°, and 75°. What is the measure of the other angle?

2. The measure of a leg of an isosceles trapezoid is 14 feet. What is the measure of the other leg?

3. The length of a diagonal of an isosceles trapezoid is 50 inches. What is the measure of the other diagonal?

QUADRILATERAL WORD PROBLEMS

Each example problem that follows uses the steps to solving word problems and the properties of quadrilaterals. Use these problems as a guide to solving quadrilateral word problems.

Example 1

An isosceles trapezoid has base angles of 120°. What is the measure in each of the other angles of the trapezoid?

Read and understand the question. This question is looking for the measure of each of the unknown angles in an isosceles trapezoid. The measures of the base angles are given.

Make a plan. Use the problem solving strategy of drawing a picture to help with this question. In an isosceles trapezoid, the two legs are congruent and the base angles are congruent. The following figure represents this trapezoid.

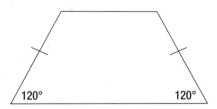

Carry out the plan. The base angles of the trapezoid are congruent, so each of their measures is 120°. To find the measure of the other angles, subtract the sum of the base angles from 360° and divide by 2: $360 - 240 = 120$, $\frac{120}{2} = 60°$ in each of the other angles. Thus, the angles measure 120°, 120°, 60°, and 60°, respectively.

Check your answer. To check this answer, add the measures of the four angles to be sure that the total number of degrees is 360: $120 + 120 + 60 + 60 = 360°$, so this answer is checking.

Example 2

A parallelogram has two opposite sides labeled $x + 5$ units and $2x - 3$ units, respectively. What is the length of these opposite sides?

Read and understand the question. This question is looking for the length of each of the opposite sides in a parallelogram.

Make a plan. The lengths of opposite sides of a parallelogram are equal. Set the given expressions equal to each other, and solve for x. Then, substitute the value of x into one of the expressions to find the length of the sides.

Carry out the plan. Set the expressions equal to each other: $x + 5 = 2x - 3$. Subtract x from each side of the equation to get $5 = x - 3$. Add 3 to each side of the equation to get $8 = x$. Substitute $x = 8$ into the expression $x + 5$. $8 + 5 = 13$. The length of each of the opposite sides is 13 units.

Check your answer. To check this answer, substitute $x = 8$ into the expression $2x - 3$ to be sure the value is also 13.

$$2(8) - 3 = 16 - 3 = 13$$

This solution is checking.

Use the steps to solving word problems and your knowledge of quadrilaterals to solve each of the questions in the following set.

PRACTICE 3

1. The angles of a quadrilateral are 65°, 95°, and 110°. What is the measure of the other angle?

2. The measure of a diagonal of a square is represented by the expression $4x - 10$. If the measure of the other diagonal is 10 meters, what is the value of x?

3. The measures of two consecutive angles of a parallelogram are repre-sented by the expressions $2x$ and $7x$, respectively. What is the value of x?

4. The opposite angles of a parallelogram are represented by the expressions $x + 18$ and $2x - 2$, respectively. What is the measure of each angle?

5. The measures of the base angles of an isosceles trapezoid are each $100°$. What is the measure of each of the other angles in the figure?

ANSWERS

Practice 1

1. Every side of a square measures the same length. Each side is 4 m.
2. In a parallelogram, the opposite angles are congruent. Thus, the value of x is $65°$.
3. In a parallelogram, the sides opposite each other are congruent. The mea-sure of the opposite side is 24 inches.
4. The measures of consecutive angles of a rhombus are supplementary, or equal to $180°$. The measure of the consecutive angle is $180 - 35 = 145°$.

Practice 2

1. To find the measure of the other angle, add the known angle measures together, and subtract the sum from $360°$. The three known angles add to $100 + 90 + 75 = 265$; $360 - 265 = 95$. The missing angle is $95°$.
2. The legs of an isosceles trapezoid have the same length. The other leg measures 14 feet.
3. The diagonals of an isosceles trapezoid are the same length. The measure of the other diagonal is 50 inches.

Practice 3

1. *Read and understand the question.* This question is looking for the missing angle of a quadrilateral when three of the angle measures are given.
   Make a plan. Add the three known angle measures, and then subtract this amount from the total of 360° in the quadrilateral.
   Carry out the plan. To find the measure of the other angle, add the known angle measures together, and subtract the sum from 360°. The three known angles add to $65 + 95 + 110 = 270$; $360 - 270 = 90$. The missing angle is 90°.
   Check your answer. To check this result, add the four angles to be sure that the sum is 360°: $65 + 95 + 110 + 90 = 360$, so this answer is checking.

2. *Read and understand the question.* This question is looking for the value of x when information is given about the diagonals of a square.
   Make a plan. The lengths of the diagonals of a square are equal. Set the given expressions equal to each other, and solve for x.
   Carry out the plan. Set the expression equal to 10: $4x - 10 = 10$. Add 10 to each side of the equation to get $4x = 20$. Divide each side of the equation by 4 to get $x = 5$.
   Check your answer. To check this answer, substitute $x = 5$ into the expression $4x - 10$ to be sure the value is 10.
   $$4(5) - 10 = 20 - 10 = 10$$
   This solution is checking.

3. *Read and understand the question.* This question asks for the value of x when expressions for two consecutive angles of a parallelogram are given.
   Make a plan. Two consecutive angles of a parallelogram have a sum of 180°; they are supplementary. Add the two expressions, set the sum equal to 180, and solve for x.
   Carry out the plan. The equation becomes $2x + 7x = 180$. Combine like terms to get $9x = 180$. Divide each side of the equation by 9 to get $x = 20$.
   Check your answer. To check this solution, substitute the value of x into each expression. Then, add the two angle measures to be sure that the total is 180°. The angles are $2(20) = 40°$ and $7(20) = 140°$. The sum of the angles is $40 + 140 = 180°$. This result is checking.

4. *Read and understand the question.* This question is looking for the measure of two opposite angles in a parallelogram.
   Make a plan. The measures of opposite angles of a parallelogram are equal. Set the given expressions equal to each other, and solve for x. Then, substitute the value of x into one of the expressions to find the measure of each angle.

Carry out the plan. Set the expressions equal to each other: $x + 18 = 2x - 2$. Subtract x from each side of the equation to get $18 = x - 2$. Add 2 to each side of the equation to get $20 = x$. Substitute $x = 20$ into the expression $x + 18$.

$$20 + 18 = 38$$

The measure of each opposite angle is 38°.

Check your answer. To check this answer, substitute $x = 20$ into the expression $2x - 2$ to be sure the value is also 38.

$$2(20) - 2 = 40 - 2 = 38$$

This solution is checking.

5. *Read and understand the question.* This question is looking for the measure of each of the unknown angles in an isosceles trapezoid. The measure of the base angles is given.

Make a plan. Use the problem solving strategy of drawing a picture to help with this question. In an isosceles trapezoid, the two legs are congruent and the base angles are congruent. The following figure represents this trapezoid.

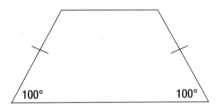

Carry out the plan. The base angles of the trapezoid are congruent, so each of their measures is 100°. To find the measure of the other angles, subtract the sum of the base angles from 360 and divide the result by 2.

$$360 - 300 = 160$$
$$\frac{160}{2} = 80$$

degrees in each of the other angles. Thus, the angles measure 100, 100, 80, and 80°, respectively.

Check your answer. To check this answer, add the measures of the four angles to be sure that the total number of degrees is 360.

$$100 + 100 + 80 + 80 = 360°$$

so this answer is checking.

similar figure and Pythagorean theorem word problems

Where there is matter, there is geometry.
—JOHANNES KEPLER (1571–1630)

This lesson will review the concepts of similar triangles and the Pythagorean theorem, and will apply these concepts to word-problem solving.

SIMILAR FIGURES

Similar figures are two or more figures that are the same shape but different sizes. The sides of similar figures are in proportion with each other and the corresponding angles are congruent.

SIMILAR TRIANGLES

There are many types of similar triangles. Some examples are shown next.

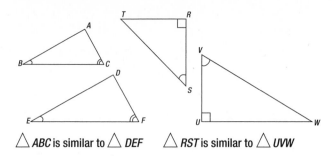

△ *ABC* is similar to △ *DEF*   △ *RST* is similar to △ *UVW*

To find the measure of missing sides of similar triangles, line up the corresponding parts in a proportion. Then, solve the proportion by cross multiplying. Take the following example.

Example

The triangles in the following figure are similar. What is the value of x in the figure?

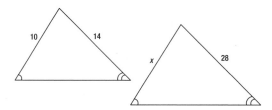

Read and understand the question. This question is looking for the value of x in a figure with two similar triangles.

Make a plan. Line up the corresponding sides in a proportion. Then, cross multiply to find the value of x.

Carry out the plan. First, identify the corresponding sides. The side labeled 10 corresponds with the side labeled x, and the side labeled 14 corresponds with the side labeled 28. Set up a proportion using the corresponding parts. Use the proportion

$$\frac{\text{side of small triangle}}{\text{side of large triangle}} = \frac{\text{side of small triangle}}{\text{side of large triangle}}$$

The proportion is

$$\frac{10}{x} = \frac{14}{28}$$

Cross multiply to get $14x = 280$. Divide each side of the equation by 14 to get $x = 20$. The value of x is 20.

Check your answer. To check this solution, substitute $x = 20$ into the proportion and cross multiply to be sure that the cross products are equal. The proportion becomes $\frac{10}{20} = \frac{14}{28}$. Cross multiply to get $280 = 280$. The cross products are equal, so this answer is checking.

OTHER SIMILAR FIGURES

Similar figures other than triangles can be solved in the same way.

Example

Two quadrilaterals are similar. The shortest side of one quadrilateral is 15 m and the longest is 25 m. If the shortest side of the other quadrilateral is 3 m, what is the measure of the longest side?

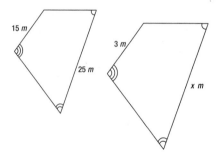

Read and understand the question. This question is looking for the value of x in a figure with two similar quadrilaterals.

Make a plan. Line up the corresponding sides in a proportion. Then, cross multiply to find the value of x.

Carry out the plan. First, identify the corresponding sides. The side that is 15 m corresponds with the side that is 3 m, and the side that is 25 m corresponds with the unknown side, or x. Set up a proportion using the corresponding parts. Use the proportion

$$\frac{\text{side of small quadrilateral}}{\text{side of large quadrilateral}} = \frac{\text{side of small quadrilateral}}{\text{side of large quadrilateral}}$$

The proportion is

$$\frac{3}{15} = \frac{x}{25}$$

Cross multiply to get $15x = 75$. Divide each side of the equation by 15 to get $x = 5$. The length of the longest side is 5 m.

Check your answer. To check this solution, substitute $x = 5$ into the proportion and cross multiply to be sure that the cross products are equal. The proportion becomes $\frac{3}{15} = \frac{5}{25}$. Cross multiply to get $75 = 75$. The cross products are equal, so this answer is checking.

....

TIP: Any similar figure problem can be solved using the preceding steps, no matter the number of sides. Simply line up the corresponding parts in a proportion and cross multiply to solve.

....

PRACTICE 1

1. A triangle has sides 4 m, 5 m, and 6 m. What is the measure of the longest side of a similar triangle whose shortest side is 16 m?

2. A triangle has sides 7 in., 10 in., and 11 in. What is the measure of the shortest side of a similar triangle whose longest side is 33 in.?

3. One rectangle has sides that are four times the size of another rectangle. If the measures of the sides of the larger rectangle are 24 cm and 32 cm, respectively, what are the measures of the sides of the smaller rectangle?

4. Two quadrilaterals are similar, and the measures of the sides of one figure are 2 m, 4 m, 5 m, and 8 m, respectively. What is the measure of the longest side of the other quadrilateral if the measure of the shortest side is 12 m?

PYTHAGOREAN THEOREM

The Pythagorean theorem is a special relationship between the sides of any right triangle. This theorem states that the sum of the square of the legs of the triangle is equal to the square of the hypotenuse. Take the following triangle labeled. The legs are labeled a and b and the hypotenuse is labeled c.

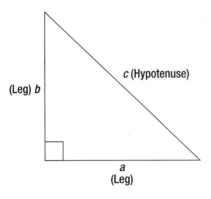

. . . .

TIP: When you are working with the sides of a right triangle, use the Pythagorean theorem, or $a^2 + b^2 = c^2$.

. . . .

Look at the following examples. In the first example, the two legs are given, and the hypotenuse is unknown. In the second example, one leg and the hypotenuse are given, and the other leg is unknown.

Example 1

The legs of a right triangle measure 3 m and 4 m. What is the measure of the hypotenuse?

Read and understand the question. This question is looking for the hypotenuse of a right triangle when the lengths of the two legs are known.

Make a plan. Use the formula $a^2 + b^2 = c^2$ and substitute the given values. Then, solve the equation for the unknown value.

Carry out the plan. The legs are 3 m and 4 m, so $a = 3$ and $b = 4$. Use the formula $a^2 + b^2 = c^2$ and substitute the given values: $3^2 + 4^2 = c^2$. Evaluate the exponents: $9 + 16 = c^2$. Add: $25 = c^2$. Take the positive square root of each side of the equation: $5 = c$. The length of the hypotenuse is 5 m.

Check your answer. To check this answer, substitute the lengths of the three sides into the formula. The formula $a^2 + b^2 = c^2$ becomes $3^2 + 4^2 = 5^2$. Apply the exponents to get $9 + 16 = 25$. Add the numbers on the left side to get $25 = 25$. This answer is checking.

Example 2

The leg of a right triangle measures 24 in. and the hypotenuse measures 26 in. What is the measure of the other leg of the triangle?

Read and understand the question. This question is looking for a leg of a right triangle when the lengths of the other leg and the hypotenuse are known.

Make a plan. Use the formula $a^2 + b^2 = c^2$ and substitute the given values. Then, solve the equation for the unknown value.

Carry out the plan. One leg is 24 in. and the hypotenuse is 26 in., so $a = 24$ and $c = 26$. Use the formula $a^2 + b^2 = c^2$ and substitute the given values: $24^2 + b^2 = 26^2$. Evaluate the exponents: $576 + b^2 = 676$. Subtract 576 from each side of the equation to get the variable alone.

$$576 - 576 + b^2 = 676 - 576$$
$$b^2 = 100$$

Take the positive square root of each side of the equation: $b = 10$. The length of the other leg is 10 in.

Check your answer. To check this answer, substitute the lengths of the three sides into the formula. The formula $a^2 + b^2 = c^2$ becomes $10^2 + 24^2 = 26^2$. Apply the exponents to get $100 + 576 = 676$. Add the numbers on the left side to get $676 = 676$. This answer is checking.

Use the Pythagorean theorem and the word-problem solving steps to solve the word problems in the following practice set.

PRACTICE 2

1. A wire attached to a 16-ft. pole is anchored into the ground 12 ft. from the base of the pole. How long is the wire?

2. A person is flying a kite. The string attached to the kite is 17 m long. If the kite is exactly 15 m horizontally from the person, how high is the kite off the ground?

3. A flagpole is 12 ft. tall. A person raising the flag is holding a rope attached to the top of the pole. If the rope is 15 ft. long, how many feet is the person from the base of the pole?

ANSWERS

Practice 1

1. *Read and understand the question.* This question is looking for the value of the longest side in a figure with two similar triangles.
   Make a plan. Use the strategy of drawing a picture. Then, line up the corresponding sides in a proportion. Finally, cross multiply to find the value of *x*, the unknown side.
   Carry out the plan. First, draw a picture. The picture of the two triangles follows.

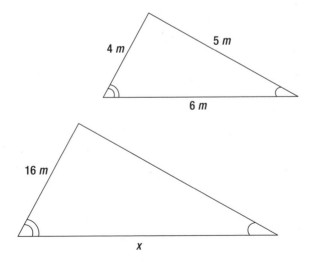

Next, identify the corresponding sides. The side labeled 6 m corresponds with the side labeled x, and the side labeled 4 m corresponds with the side labeled 16 m. Set up a proportion using the corresponding parts. Use the proportion

$$\frac{\text{side of small triangle}}{\text{side of large triangle}} = \frac{\text{side of small triangle}}{\text{side of large triangle}}$$

The proportion is

$$\frac{6}{x} = \frac{4}{16}$$

Cross multiply to get $4x = 96$. Divide each side of the equation by 4 to get $x = 24$. The value of x is 24 m.

Check your answer. To check this solution, substitute $x = 24$ into the proportion and cross multiply to be sure that the cross products are equal. The proportion becomes

$$\frac{6}{24} = \frac{4}{16}$$

Cross multiply to get $96 = 96$. The cross products are equal, so this answer is checking.

2. *Read and understand the question.* This question is looking for the value of the shortest side in a figure with two similar triangles.

Make a plan. Use the strategy of drawing a picture. Then, line up the corresponding sides in a proportion. Finally, cross multiply to find the value of x, the unknown side.

Carry out the plan. First, draw a picture. The picture of the two triangles appears next.

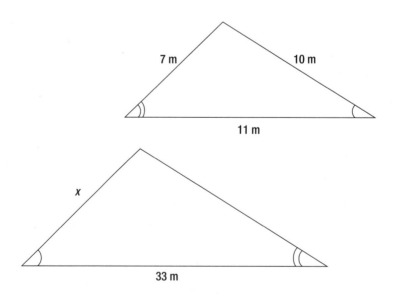

Next, identify the corresponding sides. The side labeled 7 corresponds with the side labeled x, and the side labeled 11 corresponds with the side labeled 33. Set up a proportion using the corresponding parts. Use the proportion

$$\frac{\text{side of small triangle}}{\text{side of large triangle}} = \frac{\text{side of small triangle}}{\text{side of large triangle}}$$

The proportion is

$$\frac{7}{x} = \frac{11}{33}$$

Cross multiply to get $11x = 231$. Divide each side of the equation by 11 to get $x = 21$. The value of x is 21 m.

Check your answer. To check this solution, substitute $x = 21$ into the proportion and cross multiply to be sure that the cross products are equal. The proportion becomes

$$\frac{7}{21} = \frac{11}{33}$$

Cross multiply to get $231 = 231$. The cross products are equal, so this answer is checking.

3. *Read and understand the question.* This question is looking for the lengths of the sides of a rectangle when the sides of a similar rectangle are given.

Make a plan. The sides of similar figures are in proportion. In this question, the sides of the larger rectangle are 4 times as large as the sides of a smaller rectangle. Divide the measures of these sides by 4 to find the lengths of the sides of the smaller rectangle.

Carry out the plan. The sides of the larger rectangle are 24 and 32. Twenty-four divided by 4 is 6 and 32 divided by 4 is 8. The sides of the smaller rectangle are 6 cm and 8 cm.

Check your answer. To check this solution, multiply the sides of the smaller rectangle by 4.

$$6 \times 4 = 24$$

and

$$8 \times 4 = 32$$

so this solution is checking.

4. *Read and understand the question.* This question is looking for the missing side of a quadrilateral.

Make a plan. Use the strategy of drawing a picture. Then, line up the corresponding sides in a proportion. Finally, cross multiply to find the value of x, the missing side.

Carry out the plan. First, draw a picture. The picture of the two quadrilaterals appears next.

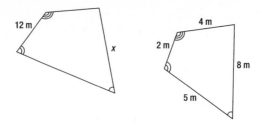

Next, identify the corresponding sides. The side that is 2 m corresponds with the side that is 12 m, and the side that is 8 m corresponds with the unknown side, or x. Set up a proportion using the corresponding parts. Use the proportion

$$\frac{\text{side of small quadrilateral}}{\text{side of large quadrilateral}} = \frac{\text{side of small quadrilateral}}{\text{side of large quadrilateral}}$$

The proportion is

$$\frac{2}{12} = \frac{8}{x}$$

Cross multiply to get $2x = 96$. Divide each side of the equation by 2 to get $x = 48$. The length of the longest side is 48 m.

Check your answer. To check this solution, substitute $x = 48$ into the proportion and cross multiply to be sure that the cross products are equal. The proportion becomes

$$\frac{2}{12} = \frac{8}{48}$$

Cross multiply to get $96 = 96$. The cross products are equal, so this solution is checking.

Practice 2

1. *Read and understand the question.* This question is looking for the length of the wire when the height of the pole and the distance from the base are known. The length of the wire is the hypotenuse of a right triangle.

   Make a plan. Use the strategy of drawing a picture. Then, use the formula $a^2 + b^2 = c^2$ and substitute the given values. Finally, solve the equation for the unknown value.

   Carry out the plan. Draw a picture of the pole and the wire, as shown in the following figure.

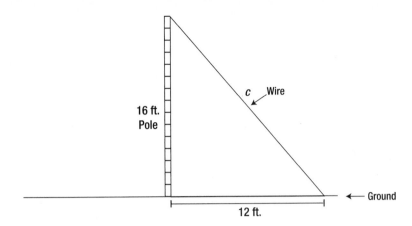

The legs are 16 ft. and 12 ft., so $a = 16$ and $b = 12$. Use the formula $a^2 + b^2 = c^2$ and substitute the given values.

$16^2 + 12^2 = c^2$

Evaluate the exponents.

$256 + 144 = c^2$

Add.

$400 = c^2$

Take the positive square root of each side of the equation.

$20 = c$

The length of the wire is 20 ft.

Check your answer. To check this answer, substitute the lengths of the three sides into the formula. The formula $a^2 + b^2 = c^2$ becomes $16^2 + 12^2 = 20^2$. Apply the exponents to get $256 + 144 = 400$. Add the numbers on the left side to get $400 = 400$. This answer is checking.

2. *Read and understand the question.* This question is looking for the height of a kite a person is flying. The length of the kite string and the horizontal distance from the person to the kite are known.

Make a plan. Use the strategy of drawing a picture. Then, use the formula $a^2 + b^2 = c^2$ and substitute the given values. Finally, solve the equation for the unknown value.

Carry out the plan. Draw a picture of the person flying the kite, as shown in the following figure.

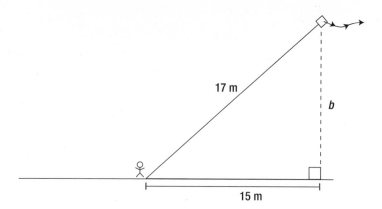

One leg is 15 m and the hypotenuse is 17 m, so $a = 15$ and $c = 17$. Use the formula $a^2 + b^2 = c^2$ and substitute the given values.

$$15^2 + b^2 = 17^2$$

Evaluate the exponents.

$$225 + b^2 = 289$$

Subtract 225 from each side of the equation to get the variable alone.

$$225 - 225 + b^2 = 289 - 225$$
$$b^2 = 64$$

Take the positive square root of each side of the equation: $b = 8$. The length of the other leg is 8, so the height of the kite is 8 m.

Check your answer. To check this answer, substitute the lengths of the three sides into the formula. The formula $a^2 + b^2 = c^2$ becomes $15^2 + 8^2 = 17^2$. Apply the exponents to get $225 + 64 = 289$. Add the numbers on the left side to get $289 = 289$. This answer is checking.

3. *Read and understand the question.* This question is looking for the distance a person is from a flagpole. The height of the pole and the length of a rope attached to the top of the pole are known.

Make a plan. Use the strategy of drawing a picture. Then, use the formula $a^2 + b^2 = c^2$ and substitute the given values. Finally, solve the equation for the unknown value.

Carry out the plan. Draw a picture of the flagpole and the person holding the rope, as shown in the following figure.

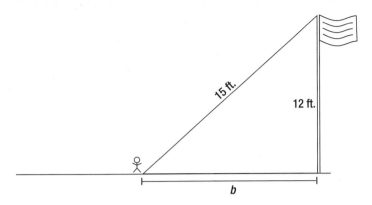

One leg is 12 ft. and the hypotenuse is 15 ft., so $a = 12$ and $c = 15$. Use the formula $a^2 + b^2 = c^2$ and substitute the given values.

$$12^2 + b^2 = 15^2$$

Evaluate the exponents.

$$144 + b^2 = 225$$

Subtract 144 from each side of the equation to get the variable alone.

$$144 - 144 + b^2 = 225 - 144$$
$$b^2 = 81$$

Take the positive square root of each side of the equation: $b = 9$. The length of the other leg is 9, so the person is 9 feet from the base of the pole.

Check your answer. To check this answer, substitute the lengths of the three sides into the formula. The formula $a^2 + b^2 = c^2$ becomes $12^2 + 9^2 = 15^2$. Apply the exponents to get $144 + 81 = 225$. Add the numbers on the left side to get $225 = 225$. This answer is checking.

perimeter word problems

Learning is not a spectator sport.
—AUTHOR UNKNOWN

This lesson reviews perimeter and the steps to solving word problems involving perimeter.

PERIMETER OF FIGURES

The perimeter of a figure is the distance around the outside of the figure. To find the perimeter, add the lengths of each of the sides together to get the total distance around the object.

Find the perimeter of each of the following figures.

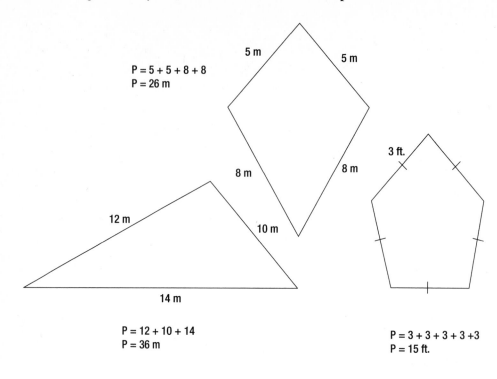

$$P = 5 + 5 + 8 + 8$$
$$P = 26 \text{ m}$$

5 m

5 m

8 m

8 m

3 ft.

12 m

10 m

14 m

$$P = 12 + 10 + 14$$
$$P = 36 \text{ m}$$

$$P = 3 + 3 + 3 + 3 + 3$$
$$P = 15 \text{ ft.}$$

TIP: The **perimeter** of any object is the distance around the object. Instead of memorizing different formulas, just add the sides to find the perimeter of an object.

TRIANGLES

Every triangle has three sides. To find the perimeter of triangles, add the lengths of the three sides together. If each side is not given, use the clues in the problem to find the measure of the sides.

Example 1

An equilateral triangle has a side of 14 meters. What is the perimeter of the triangle?

Read and understand the question. This question is looking for the perimeter of a triangle.

Make a plan. The perimeter is the distance around the object. Add the lengths of the three sides of the triangle to find the perimeter. Since the triangle is equilateral, each of the three sides is 14 meters.

Carry out the plan. Add the three sides: $14 + 14 + 14 = 42$. The perimeter is 42 meters.

Check your answer. To check this solution, start with the perimeter of 42 meters. If each side length is subtracted from this value, the result should be zero: $42 - 14 - 14 - 14 = 0$, so this solution is checking.

Example 2

An isosceles triangle has sides 10 meters, 10 meters, and 12 meters. What is the perimeter?

Read and understand the question. This question is looking for the perimeter of the triangle.

Make a plan. The perimeter is the distance around the object. Add the lengths of the three sides of the triangle to find the perimeter. All three side lengths are given.

Carry out the plan. Add the three sides: $10 + 10 + 12 = 32$ meters. The perimeter is 32 meters.

Check your answer. To check this solution, start with the perimeter of 32 meters. If each side length is subtracted from this value, the result should be zero: $32 - 10 - 10 - 12 = 0$, so this solution is checking.

RECTANGLES

Each rectangle has four sides and the opposite sides are congruent. Here is an example using the steps to word-problem solving to find the perimeter of a rectangle.

Example

The length of a rectangle is 24 inches and the width is 13 inches. What is the perimeter?

Read and understand the question. This question is looking for the perimeter of the rectangle.

Make a plan. To find the perimeter, find the distance around the object. Because it is a rectangle, the opposite sides are equal. The four sides are 24 inches, 24 inches, 13 inches, and 13 inches.

Carry out the plan. Add the sides to find the perimeter: $24 + 24 + 13 + 13 = 74$. The perimeter is 74 inches.

Check your answer. To check this solution, start with the perimeter of 74 inches. If each side length is subtracted from this value, the result should be zero: $74 - 24 - 24 - 13 - 13 = 0$, so this solution is checking.

PARALLELOGRAMS

Parallelograms are also quadrilaterals. Like rectangles, their opposite sides are congruent. The following examples model different types of parallelogram perimeter problems.

Example 1

The dimensions of a parallelogram are given next.

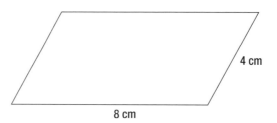

8 cm

4 cm

What is the perimeter of the parallelogram?

Read and understand the question. This question is looking for the perimeter of the parallelogram.

Make a plan. Two of the dimensions are given in the diagram. Because the figure is a parallelogram, the opposite sides are congruent. Thus, the four sides measure 4 cm, 4 cm, 8 cm, and 8 cm.

Carry out the plan. Add the sides: $4 + 4 + 8 + 8 = 24$. The perimeter is 24 cm.

Check your answer. To check this solution, start with the perimeter of 24 cm. If each side length is subtracted from this value, the result should be zero: $24 - 4 - 4 - 8 - 8 = 0$, so this solution is checking.

Example 2

The base of a parallelogram is four times the length of the adjacent side. If the perimeter is 120 meters, what is the length of the side adjacent to the base of the parallelogram?

Read and understand the question. This question is looking for the length of the side adjacent, or next to, the base of the parallelogram. The relationship between this side and the base is given, along with the perimeter of the parallelogram.

Make a plan. Use the relationship between the base and the adjacent side to write *let* statements. Then, add the four sides together and set the equation equal to 120.

Carry out the plan. Let x = adjacent side, and let $4x$ = base. The perimeter is $x + 4x + x + 4x = 120$. Simplify to get $10x = 120$. Divide each side by 10 to get $x = 12$. The length of the side is 12 meters.

Check your answer. To check your solution, substitute $x = 12$ into the perimeter formula to make sure that the perimeter is equal to 120 meters. Since $x = 12$, $4x = 4(12) = 48$ meters. Add the four sides: $12 + 48 + 12 + 48 = 120$ meters. This solution is checking.

SQUARES

Squares are quadrilaterals with four equal sides. The following example solves for the perimeter of a square.

Example

A side of a square measures 7 cm. What is the perimeter?

Read and understand the question. This question is looking for the perimeter of a square.

Make a plan. The perimeter is the distance around the object. Since the figure is a square, the sides are of equal length.

Carry out the plan. Add the sides: $7 + 7 + 7 + 7 = 28$. The perimeter is 28 cm.

Check your answer. To check this solution, start with the perimeter of 28 cm. If each side length is subtracted from this value, the result should be zero: $28 - 7 - 7 - 7 - 7 = 0$, so this solution is checking.

PERIMETER OF REGULAR POLYGONS

....

TIP: A **regular polygon** is a figure whose sides are the same lengths and whose angles are all the same measures.

....

The perimeter of a regular hexagon is 24 units. What is the measure of each side of the figure?

Read and understand the question. This question is looking for the perimeter of a regular hexagon.

Make a plan. The perimeter is the distance around an object. A regular hexagon is a six-sided polygon where each side has the same measure.

Carry out the plan. Since we are given the perimeter of a hexagon, and a hexagon has six sides, divide the perimeter by six to find the length of one side: $\frac{24}{6} = 4$ units.

Check your answer. To check this solution, start with the perimeter of 24 units. If each side length is subtracted from this value, the result should be zero: $24 - 4 - 4 - 4 - 4 - 4 - 4 = 0$, so this solution is checking.

PRACTICE

1. The measure of the base of an isosceles triangle is 10 units. If the legs each measure 2 units more than the base, what is the perimeter of the triangle?

2. The sides of a scalene triangle are 12, 15, and 20 units, respectively. What is the perimeter of the triangle?

3. A side of a square is represented by the expression $5x$. If the perimeter is 40 units, what is the length of one side of the square?

4. The length of a rectangle is three times its width. If the perimeter of the rectangle is 80 meters, what is the length of the rectangle?

5. An area is enclosed with a fence in the shape of a regular pentagon. If each side of fencing is 35 meters long, what is the perimeter of the pentagon?

6. A regular octagon has sides that each measure 20 cm. What is the perimeter of the octagon?

ANSWERS

Practice

1. *Read and understand the question.* This question is looking for the perimeter of an isosceles triangle.
   Make a plan. The perimeter is the distance around the object. Add the lengths of the three sides of the triangle to find the perimeter. The length of the base is given.
   Carry out the plan. The lengths of the two legs of the triangle are two more than the length of the base side: $10 + 2 = 12$. Add the three sides: $10 + 12 + 12 = 34$ meters. The perimeter is 34 meters.
   Check your answer. To check this solution, start with the perimeter of 34 meters. If each side length is subtracted from this value, the result should be zero: $34 - 10 - 12 - 12 = 0$, so this solution is checking.

2. *Read and understand the question.* This question is looking for the perimeter of a scalene triangle.
   Make a plan. The perimeter is the distance around the object. Add the lengths of the three sides of the triangle to find the perimeter. All three side lengths are given.
   Carry out the plan. Add the three side lengths: $12 + 15 + 20 = 47$ meters. The perimeter is 47 meters.
   Check your answer. To check this solution, start with the perimeter of 47 meters. If each side length is subtracted from this value, the result should be zero: $47 - 12 - 15 - 20 = 0$, so this solution is checking.

3. *Read and understand the question.* This question is looking for the side of a square when an expression for the sides and the perimeter is known.
   Make a plan. The perimeter is the distance around the object. Since the figure is a square, each of the sides are of equal length. Use $5x$ for the length of each of the four sides.
   Carry out the plan. Add the expression that represents each side, and set the value equal to 40. Then solve for x.
   $$5x + 5x + 5x + 5x = 40$$
   Combine like terms.
   $$20x = 40$$
   Divide each side by 20 to get $x = 2$. Substitute $x = 2$ into $5x$ to get the length of one side.
   $$5(2) = 10$$

The length of one side is 10 units. Another way to solve this question is to use that fact that the figure is a square and the perimeter is 40. Divide 40 by 4 to get a side length of 10 units.

Check your answer. To check this solution, add the four sides to be sure the perimeter is 40.

$$10 + 10 + 10 + 10 = 40$$

so this solution is checking.

4. *Read and understand the question.* This question is looking for the length of the rectangle when the relationship between the length and the width is given, along with the perimeter.

Make a plan. Use the relationship between the length and width to write let statements. Then, add the four sides together and set the equation equal to 80.

Carry out the plan. The length is three times the width, so let w = width, and let 3w = length. The perimeter is w + 3w + w + 3w = 80. Simplify to get 8w = 80. Divide each side by 8 to get w = 10. The length is 3(10) = 30 meters.

Check your answer. To check your solution, substitute w = 10 into the perimeter formula to make sure that the perimeter is equal to 80 meters. Since w = 10, 3w = 3(10) = 30. Adding the four sides becomes 10 + 30 + 10 + 30 = 80. This answer is checking.

5. *Read and understand the question.* This question is looking for the perimeter of a regular pentagon.

Make a plan. The perimeter is the distance around an object. A regular pentagon is a five-sided polygon where each side has the same measure. Each side of the fence is 35 meters long.

Carry out the plan. Add the sides: 35 + 35 + 35 + 35 + 35, or 5 × 35 = 175. The perimeter is 175 meters.

Check your answer. To check this solution, start with the perimeter of 175 meters. If each side length is subtracted from this value, the result should be zero.

$$175 - 35 - 35 - 35 - 35 - 35 = 0$$

so this solution is checking.

6. *Read and understand the question.* This question is looking for the perimeter of a regular octagon.

Make a plan. The perimeter is the distance around an object. A regular octagon is an eight-sided polygon where each side has the same measure. Each side is 20 cm.

Carry out the plan. Add the sides: $20 + 20 + 20 + 20 + 20 + 20 + 20 + 20$, or $8 \times 20 = 160$. The perimeter is 160 cm.

Check your answer. To check this solution, start with the perimeter of 160 units. If each side length is subtracted from this value, the result should be zero.

$$160 - 20 - 20 - 20 - 20 - 20 - 20 - 20 - 20 = 0$$

so this solution is checking.

circle word problems

Numbers are intellectual witnesses that belong only to mankind.
—HONORE DE BALZAC (1799–1850)

This lesson includes circle word problems while reviewing the basic parts and formulas involved with circles.

TIP: Remember that a **circle** is the set of points that is an equal distance from a single point. The circle is named by the center point. The total number of degrees in any circle is 360.

BASIC PARTS OF CIRCLES

There are some basic parts of the circle that are necessary to know when you are applying the formulas related to circles. Each of these important parts of circle B is shown in the following figure. The point B is the center of the circle.

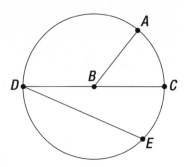

Radius: The distance from the center of a circle to a point on the circle. The plural of radius is radii. The radii in the figure are $\overline{BA}$, $\overline{BC}$, and $\overline{BD}$.

Chord: A line segment whose endpoints are on the circle. Chords $\overline{DC}$ and $\overline{DE}$ are shown in the figure.

Diameter: A chord that goes through the center of a circle. The diameter $\overline{DC}$ is shown in the figure.

Arc: A section of the circle. The symbol for an arc is a curved line above the letters of the arc. Arc $\overset{\frown}{AC}$ is shown in the figure.

Central Angle: An angle whose vertex is the center of the circle. The measure of the central angle is the same as the measure of the arc it intercepts. The angle $\angle ABC$ in the figure is a central angle.

Inscribed Angle: An angle whose vertex is on the circle. The measure of an inscribed angle is equal to half of the arc it intercepts. The angle $\angle CDE$ in the figure is an inscribed angle.

PRACTICE 1

Use the following figure to identify the basic parts of the circle.

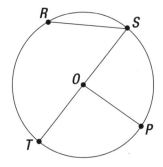

Center   _____
Radius   _____
Diameter   _____
Central Angle   _____
Inscribed Angle _____

QUESTIONS ABOUT SPECIAL ANGLES AND CIRCLES

There are some special angles that are located within circles and are related to the center of the circle and the chords of the circle. Two of these special angles are central angles and inscribed angles.

Example 1

In this figure the measure of central angle $\angle ABC$ is 30°. What is the measure of the intercepted arc $\overarc{AC}$?

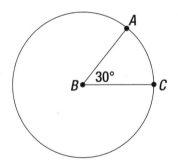

Read and understand the question. This question is looking for the measure of the arc intercepted by a central angle.

Make a plan. The measure of an arc is equal to the measure of the central angle that intercepts it. Find the measure of this angle to find the measure of the arc.

Carry out the plan. The measure of the central angle is 30°, so the measure of the arc is also 30°.

Check your answer. The measure of the arc is the same number of degrees as the central angle. This result is checking.

Example 2

In this figure the measure of inscribed angle ∠*AOB* is 20°. What is the measure of the intercepted arc $\overarc{AB}$?

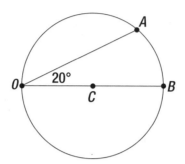

Read and understand the question. This question is looking for the measure of the arc intercepted by an inscribed angle.

Make a plan. The measure of the arc is equal to twice the measure of the inscribed angle that intercepts it. Find the measure of this angle, and multiply by two to find the measure of the arc.

Carry out the plan. The inscribed angle measures 20°. Multiply $2 \times 20 = 40$. The intercepted arc measures 40°.

Check your answer. To check the solution, divide the result by 2 to find the measure of the inscribed angle: $40 \div 2 = 20°$ in the inscribed angle. This answer is checking.

....

TIP: Remember that the measure of a **central angle** is equal to the measure of the intercepted arc, and the measure of an **inscribed angle** is half the measure of the intercepted arc.

....

PRACTICE 2

1. The measure of a central angle is 50°. What is the measure of the arc it intercepts?

2. The measure of an arc on a circle is 25°. What is the measure of the central angle that intercepts it?

3. The measure of an inscribed angle is 40°. What is the measure of the arc on the circle that it intercepts?

4. The measure of an arc on a circle is 34°. What is the measure of the inscribed angle that intercepts it?

CIRCUMFERENCE

Just like perimeter, the circumference of a circle is the distance around the circle. But because a circle does not have sides to add to find the distance around, there is a different formula. Use the formula *Circumference* $= \pi \times diameter$, or $C = \pi d$.

Example 1

What is the circumference of a circle with a diameter of 5 cm? (Leave your answer in terms of π.)

Read and understand the question. This question is asking for the circumference of a circle when the length of the diameter is given.

Make a plan. Use the formula $C = \pi d$ and substitute $d = 5$. Leave the symbol π in your answer as stated in the question.

Carry out the plan. The formula $C = \pi d$ becomes $C = \pi(5)$. The circumference is 5π cm.

Check your answer. To check your answer, substitute the circumference into the formula and make sure that the two sides are equivalent. The formula is $5\pi = \pi(5)$, so this problem is checking.

Example 2

The circumference of a circle is equal to 14π units. What is the length of the radius of the circle?

Read and understand the question. This question is looking for the length of the radius of a circle when the circumference is given.

Make a plan. Use the formula $C = \pi d$, and substitute the given circumference. Then, divide each side of the equation by π to find the length of the diameter. Take half of this value to find the length of the radius.

Carry out the plan. The formula $C = \pi d$ becomes $14\pi = \pi d$. Thus, $d = 14$. Take half of this value to find the radius. The radius is equal to 7 units.

Check your answer. To check this solution, find the diameter by multiplying the radius by 2, and substitute this value into the formula for d. The diameter is equal to $7 \times 2 = 14$, so the circumference is $C = \pi(14) = 14\pi$. This result is checking.

PRACTICE 3

1. The diameter of a circle is 16 cm. What is the circumference of the circle in terms of π?

2. The radius of a circle is 6 m. What is the circumference of the circle in terms of π?

3. The circumference of a circle is 30π cm. What is the diameter of the circle?

4. The circumference of a circle is 100π cm. What is the radius of the circle?

ANSWERS

Practice 1

Center	_____O_____
Radius	$\overline{OT}$, $\overline{OS}$, or $\overline{OP}$
Diameter	_____ $\overline{TS}$_____
Central Angle	_____$\angle SOP$_____
Inscribed Angle	_____$\angle RST$_____

Practice 2

1. *Read and understand the question.* This question is looking for the measure of the arc intercepted by a central angle.
   Make a plan. The measure of an arc is equal to the measure of the central angle that intercepts it. Find the measure of this angle to find the measure of the arc.
   Carry out the plan. The measure of the central angle is 50°, so the measure of the arc is also 50°.
   Check your answer. The measure of the arc is the same number of degrees as the central angle. This result is checking.

2. *Read and understand the question.* This question is looking for the measure of the central angle when the measure of the intercepted arc is known.
   Make a plan. The measure of an arc is equal to the measure of the central angle that intercepts it. Use the measure of the arc to find the measure of the central angle.
   Carry out the plan. The measure of the arc is 25°, so the measure of the central angle is also 25°.
   Check your answer. The measure of the arc is the same number of degrees as the central angle. This result is checking.

3. *Read and understand the question.* This question is looking for the measure of the arc intercepted by an inscribed angle.
   Make a plan. The measure of the arc is equal to twice the measure of the inscribed angle that intercepts it. Find the measure of this angle and multiply by 2 to find the measure of the arc.
   Carry out the plan. The inscribed angle measures 40°. Multiply $2 \times 40 = 80$. The intercepted arc measures 80°.
   Check your answer. To check the solution, divide the result by 2 to find the measure of the inscribed angle: $80 \div 2 = 40°$ in the inscribed angle. This answer is checking.

4. *Read and understand the question.* This question is looking for the measure of the inscribed angle when the measure of the intercepted arc is known.
   Make a plan. The measure of the inscribed angle is half the measure of the arc it intercepts. Find the measure of this angle by dividing the measure of the arc by 2.
   Carry out the plan. The intercepted arc measures 34°. Divide $34 \div 2 = 17$. The inscribed angle measures 17°.
   Check your answer. To check the solution, multiply the result by 2 to find the measure of the intercepted arc: $17 \times 2 = 34°$ in the intercepted arc. This answer is checking.

Practice 3

1. *Read and understand the question.* This question is asking for the circumference of a circle when the length of the diameter is given.

   Make a plan. Use the formula $C = \pi d$ and substitute $d = 16$. Leave the symbol π in your answer as stated in the question.

   Carry out the plan. The formula $C = \pi d$ becomes $C = \pi(16)$. The circumference is 16π cm.

   Check your answer. To check your answer, substitute the circumference into the formula, and make sure that the two sides are equivalent. The formula is $16\pi = \pi(16)$, so this problem is checking.

2. *Read and understand the question.* This question is asking for the circumference of a circle when the length of the radius is given.

   Make a plan. Multiply the value of the radius by 2 to find the diameter: $6 \times 2 = 12$. Use the formula $C = \pi d$ and substitute $d = 12$. Leave the symbol π in your answer as stated in the question.

   Carry out the plan. The formula $C = \pi d$ becomes $C = \pi(12)$. The circumference is 12π m.

   Check your answer. To check your answer, substitute the circumference into the formula and make sure that the two sides are equivalent. The formula is $12\pi = \pi(12)$. The diameter is 12 m, so the radius is equal to 6 m. This answer is checking.

3. *Read and understand the question.* This question is looking for the length of the diameter of a circle when the circumference is given.

   Make a plan. Use the formula $C = \pi d$ and substitute the given circumference. Then, divide each side of the equation by π to find the length of the diameter.

   Carry out the plan. The formula $C = \pi d$ becomes $30\pi = \pi d$. Thus, $d = 30$. The diameter is equal to 30 cm.

   Check your answer. To check this solution, substitute the value of the diameter into the formula for d. The diameter is equal to 30 cm, so the circumference is $C = \pi(30) = 30\pi$ cm. This result is checking.

4. *Read and understand the question.* This question is looking for the length of the radius of a circle when the circumference is given.

Make a plan. Use the formula $C = \pi d$ and substitute the given circumference. Then, divide each side of the equation by π to find the length of the diameter. Take half of this value to find the length of the radius.

Carry out the plan. The formula $C = \pi d$ becomes $100\pi = \pi d$. Thus, $d = 100$. Take half of this value to find the radius. The radius is equal to 50 cm.

Check your answer. To check this solution, find the diameter by multiplying the radius by 2, and substitute this value into the formula for d. The diameter is equal to $50 \times 2 = 100$, so the circumference is $C = \pi(100) = 100\pi$ cm. This result is checking.

area word problems

*One cannot escape the feeling that these mathematical formulas
have an independent existence and an intelligence
of their own, that they are wiser than we are,
wiser even than their discoverers. . .*
—HEINRICH HERTZ (1857–1894)

This lesson reviews the area formulas and their usefulness in solving word problems.

AREA FORMULAS

Area is the number of square units that can cover a certain region. The area of various shapes in geometry can be calculated by using different formulas.

TIP: The **area** of many geometric figures is determined by the general formula *Area = base × height*, or $A = bh$. Area is always expressed in square units.

Parallelograms

The formula for finding the area of any shape in the parallelogram group of figures is *Area = base × height*, or $A = b \times h$. The shapes with these properties also include rectangles, rhombuses, and squares.

Example 1
What is the area of a parallelogram with a base of 8 cm and a height of 7 cm?

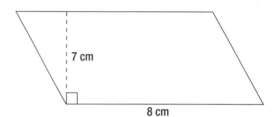

Read and understand the question. This question is looking for the area of a parallelogram when the base and height are given.

Make a plan. Use the formula $A = b \times h$ and substitute the given values to find the area.

Carry out the plan. Substitute the given values into the formula to get $A = 8 \times 7 = 56$ cm^2.

Check your answer. To check the solution, divide the area by one of the dimensions and make sure the result is the other dimension: $56 \div 8 = 7$. This answer is checking.

Example 2
What is the base of a rectangle with an area of 48 m^2 and a height of 6 m?

Read and understand the question. This question is looking for the base of a rectangle when the area and the height are given.

Make a plan. Use the formula $A = b \times h$ and substitute the given values to solve for the base.

Carry out the plan. Substitute the given values into the formula to get $48 = b \times 6$. Divide each side of the equation by 6 to get $b = 8$ m.

Check your answer. To check the solution, divide the area by one of the dimensions and make sure the result is the other dimension. $48 \div 6 = 8$. This answer is checking.

Squares

The area of a square can be found by using the formula for parallelograms, but the area can also be expressed by the formula *Area = side × side*, or $A = s^2$.

Example
The length of a side of a square is 9 in. What is the area of the square?

Read and understand the question. This question is looking for the area of a square when the length of a side is known.

Make a plan. Use the formula $A = b \times h$ or s^2, and substitute the given value of the side.

Carry out the plan. Substitute the given values into the formula to get $A = 9 \times 9 = 81$ in.^{2}

Check your answer. To check the solution, divide the area by the length of a side and make sure the result is also the measure of a side: $81 \div 9 = 9$. This answer is checking.

Triangles

The area of any triangle is equal to half of the area of a parallelogram with the same base and height. Because of this, the area of any triangle can be found by using the formula *Area = one-half of the base × height*, or $A = \frac{1}{2}bh$.

Example
The length of the base of a triangle is 10 units and the height is 12 units. What is the number of square units in the area?

Read and understand the question. This question is looking for the area of a triangle when the base and height are given.

Make a plan. Use the formula $A = \frac{1}{2}bh$ and substitute the given values to find the area.

Carry out the plan. Substitute the given values into the formula to get $A = \frac{1}{2}(10)(12) = \frac{1}{2}(120) = 60^2$ units.

Check your answer. To check the solution, work backward by doubling the area and then dividing this result by one of the dimensions to make sure the result is the other dimension.

$$2 \times 60 = 120$$
$$120 \div 10 = 12$$

This answer is checking.

Circles

The area of a circle can be found by using the formula *Area = the square of the radius* $\times \pi$, or $A = \pi r.^2$

Example 1
What is the area of a circle with a radius of 4 m? Leave the answer in terms of π.

Read and understand the question. This question asks for the area of a circle when the radius is known.

Make a plan. Use the formula $A = \pi r^2$ and substitute the given value for the radius.

Carry out the plan. Substitute into the formula to get $A = \pi \times 4^2 = 16\pi$ m^2.

Check your answer. To check the solution, divide the area by π times the radius. The result is 4, which is the length of the radius. This answer is checking.

Example 2
What is the length of the diameter of a circle with an area of 121π cm^2?

Read and understand the question. This question asks for the diameter of a circle when the area is known.

Make a plan. Use the formula $A = \pi r^2$ and substitute the given value for the area. Then, solve the equation for r and multiply the result by 2 to find the diameter.

Carry out the plan. Substitute into the formula to get $121\pi = \pi \times r^2$. Divide each side of the equation by π to get $r^2 = 121$. Take the positive square root of each side of the equation to get $r = 11$. Multiply this result by 2 to find the length of the diameter.

$2 \times 11 = 22$ cm

Check your answer. To check the solution, divide the diameter by 2 and substitute this value into the area formula.

$22 \div 2 = 11$
$A = \pi(11)^2 = 121\pi$ cm^2

This answer is checking.

PRACTICE

1. What is the area of a rectangle with a base of 6 cm and a height of 5 cm?

2. The base of a triangle is 12 m and the height is 14 m. What is the area of the triangle?

3. The area of a parallelogram is 52 cm^2 and the height is 13 cm. What is the base of the parallelogram?

4. A square backyard has an area of 144 square feet. What is the length of a side of the backyard?

5. A circle has a radius of 7 m. What is the area of the circle?

6. The area of a circular animal pen is 196π square feet. What is the diameter of the pen?

ANSWERS

Practice 1

1. *Read and understand the question.* This question is looking for the area of a parallelogram when the base and height are given.
   Make a plan. Use the formula $A = b \times h$ and substitute the given values to find the area.
   Carry out the plan. Substitute the given values into the formula to get $A = 6 \times 5 = 30$ cm^2.
   Check your answer. To check the solution, divide the area by one of the dimensions and make sure the result is the other dimension: $30 \div 6 = 5$. This answer is checking.

2. *Read and understand the question.* This question is looking for the area of a triangle when the base and height are given.
   Make a plan. Use the formula $A = \frac{1}{2}bh$ and substitute the given values to find the area.
   Carry out the plan. Substitute the given values into the formula to get $A = \frac{1}{2}(12)(14) = \frac{1}{2}(168) = 84$ m^2.
   Check your answer. To check the solution, work backward by doubling the area and then dividing this result by one of the dimensions to make sure the result is the other dimension.

$$2 \times 84 = 168$$
$$168 \div 12 = 14$$

   This answer is checking.

3. *Read and understand the question.* This question is looking for the base of a parallelogram when the area and height are given.
   Make a plan. Use the formula $A = b \times h$ and substitute the given values to solve for the base.
   Carry out the plan. Substitute the given values into the formula to get $52 = b \times 13$. Divide each side of the equation by 13 to get $b = 4$ cm.
   Check your answer. To check the solution, divide the area by one of the dimensions, and make sure the result is the other dimension: $52 \div 4 = 13$. This answer is checking.

4. *Read and understand the question.* This question is looking for the length of the side of a square backyard when the area is known.
   Make a plan. Use the formula $A = b \times h$ or s^2, and substitute the given value of the side.
   Carry out the plan. Substitute the given values into the formula to get $144 = s^2$. Take the positive square root of each side of the equation to get $s = 12$ ft.

Check your answer. To check the solution, square the length of a side and make sure the result is the area: $12 \times 12 = 144$. This answer is checking.

5. *Read and understand the question.* This question asks for the area of a circle when the radius is known.

   Make a plan. Use the formula $A = \pi r^2$ and substitute the given value for the radius.

   Carry out the plan. Substitute into the formula to get $A = \pi \times 7^2 = 49\pi$ m^2.

   Check your answer. To check the solution, divide the area by π times the radius. The result is 7, which is the length of the radius. This answer is checking.

6. *Read and understand the question.* This question asks for the diameter of a circle when the area is known.

   Make a plan. Use the formula $A = \pi r^2$ and substitute the given value for the area. Then, solve the equation for r and multiply the result by 2 to find the diameter.

   Carry out the plan. Substitute into the formula to get $196\pi = \pi \times r^2$. Divide each side of the equation by π to get $r^2 = 196$. Take the positive square root of each side of the equation to get $r = 14$. Multiply this result by 2 to find the length of the diameter: $2 \times 14 = 28$ ft.

   Check your answer. To check the solution, divide the diameter by 2 and substitute this value into the area formula.

   $$28 \div 2 = 14$$
   $$A = \pi(14)^2 = 196\pi \text{ ft}^2$$

   This answer is checking.

surface area word problems

Physics is mathematical not because we know so much about the physical world, but because we know so little; it is only its mathematical properties that we can discover.

—BERTRAND RUSSELL (1872–1970)

This chapter will include details on common geometric solids and the use of the formulas needed to find the surface area of each figure. Strategies on solving word problems with surface area will also be modeled and explained.

GEOMETRIC SOLIDS

Geometric solids are three-dimensional figures that are defined by the number of faces, edges, and vertices that they have.

....

TIP: Here are some important terms to know about geometric solids.

Face: the two-dimensional "sides" of a figure
Edge: the line segment formed where two faces meet
Vertices: the points where two or more edges meet

....

The **surface area** of a three-dimensional figure is the number of square units it takes to cover each face of the figure. You can think of it as the outer cover of the object.

RECTANGULAR PRISMS

A rectangular prism is a geometric solid that has a rectangle for each of the faces. There are three dimensions in this solid: a length, a width, and a height. This is shown in the following figure.

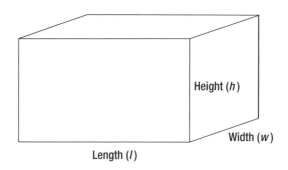

Length (*l*)

In a rectangular prism, each of the opposite faces is congruent. To find the surface area, find the area of three of the rectangular faces and multiply each by two to represent the side opposite. Then, add these values together. The formula is $SA = 2lw + 2lh + 2wh$ where w is the width, l is the length, and h is the height of the prism.

Example

What is the volume of a rectangular prism with a length of 5 m, a width of 6 m, and a height of 10 m?

Read and understand the question. This question is looking for the surface area of a rectangular prism. Each of the three dimensions is known.

Make a plan. Use the formula $SA = 2lw + 2lh + 2wh$, and substitute the given values for the length, width, and height.

Carry out the plan. The formula becomes $SA = 2(5)(6) + 2(5)(10) + 2(6)(10)$. Multiply to get $SA = 60 + 100 + 120$. Add to get a surface area of 280 m².

Check your answer. Substitute the values into the formula again to double-check your solution. The formula $SA = 2(5)(6) + 2(5)(10) + 2(6)(10)$ simplifies to $2(30) + 2(50) + 2(60) = 60 + 100 + 120 = 280$. This answer is checking.

A SPECIAL PRISM: CUBES

A cube is a special type of rectangular prism. The faces of a cube are six congruent squares, as shown in the following figure.

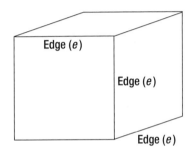

The area of each face can be found by multiplying the length of each edge of the cube by itself. To find the surface area of a cube, find the area of one face and multiply this area by six to represent all of the faces. Therefore, the formula is $SA = 6e^2$, where e is the length of an edge of the cube.

Example 1

What is the surface area of a cube with an edge that measures 7 in.?

Read and understand the question. This question is looking for the surface area of a cube when the measure of the edge of the cube is given.

Make a plan. Use the formula $SA = 6e^2$ and substitute the value of e. Remember that the surface area will be represented in square units.

Carry out the plan. Substitute into the formula to get $SA = 6(7)^2$. Evaluate the exponent first to get $SA = 6(49)$. Multiply to find the surface area: $SA = 294$ in.^{2}.

Check your answer. To check your answer, divide the total surface area by 6: $294 \div 6 = 49$. Then, take the positive square root of 49 to find the measure of the edge of the cube. The positive square root of 49 is 7, so this result is checking.

Example 2

What is the measure of the edge of a cube with a total surface area of 54 ft.²?

Read and understand the question. This question is looking for the measure of the edge of a cube when the total surface area is known.

Make a plan. Use the formula for the surface area of a cube, and work backward to solve for *e*.

Carry out the plan. Substitute the values into the formula to get $54 = 6e^2$. Divide each side of the equation by 6 to get $9 = e^2$. Take the positive square root of each side of the equation to find the value of *e*: $e = 3$ ft.

Check your answer. To check this answer, substitute the value of *e* into the surface area formula. The formula becomes $SA = 6(3)^2$. Evaluate the exponent to get $SA = 6(9)$. Multiply to get a surface area of 54 ft.². This result is checking.

PRACTICE 1

1. What is the surface area of a rectangular prism with a length of 15 cm, a height of 12 cm, and a width of 11 cm?

2. What is the surface area of a rectangular prism with a length of 20 m, a height of 21 m, and a width of 4 m?

3. What is the surface area of a cube with an edge of 9 in.?

4. The surface area of a cube is 96 m². What is the length of an edge of the cube?

SOLIDS WITH CURVED SURFACES

Cylinders

A cylinder is a solid with two circles as the bases. A right cylinder has the height perpendicular to the bases; in other words, the cylinder is upright and not tilted at all.

To find the surface area of a cylinder, imagine a label on a soup can. If the label is peeled off and flattened, it is in the shape of a rectangle. The height of this rectangle is the height of the cylinder, and the base of this rectangle is the circumference of the base of the cylinder. (See the following figure.) The area of this rectangle can be expressed as πdh. The area of each of the circular bases can be found by the expression πr^2, the formula for the area of a circle.

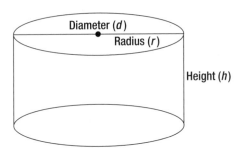

Thus, the formula for the surface area of a cylinder is $SA = 2\pi r^2 + \pi dh$, where r is the radius of the base, d is the diameter of the base, and h is the height of the cylinder.

Example

What is the surface area of a cylinder with a height of 8 m and a base with a radius of 6 m? (Leave your answer in terms of π.)

Read and understand the question. This question asks for the surface area of a cylinder when the dimensions are given.

Make a plan. Use the formula for the surface area of a cylinder, and substitute the known values. Evaluate to find the surface area.

Carry out the plan. Because the radius is 6 m, the diameter is 2×6, or 12 m. The formula $SA = 2\pi r^2 + \pi dh$ becomes $SA = 2\pi(6)^2 + \pi(12)(8)$. Evaluate the exponents and multiply to get $SA = 72\pi + 96\pi$. Combine like terms to get $SA = 168\pi$ m^2.

Check your answer. Substitute again into the formula to double-check your solution. The formula is

$$SA = 2\pi(6)^2 + \pi(12)(8) = 2\pi(36) + 96\pi = 72\pi + 96\pi = 168\pi$$

This answer is checking.

Spheres

A sphere is a round three-dimensional figure in the shape of a ball. An example of a sphere is shown in the following figure.

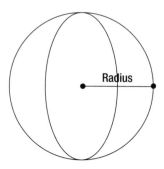

Think about having to wrap a baseball in paper. The amount of paper needed would be approximately equal to four times the area of a circle with the same radius. Therefore, the formula for the surface area of a sphere is $SA = 4\pi r^2$, where r is the length of the radius of the sphere.

Example
What is the surface area of a sphere with a radius of 3 m? (Leave your answer in terms of π.)

Read and understand the question. This question is looking for the surface area of a sphere when the radius of the sphere is known.

Make a plan. Use the formula $SA = 4\pi r^2$, and substitute $r = 3$.

Carry out the plan. The formula becomes $SA = 4\pi(3)^2$. Evaluate the exponent to get $SA = 4\pi(9)$. Multiply to get the total surface area: $SA = 36\pi$ m².

Check your answer. To check this solution, work backward, and divide the total surface area by 4π: 36π divided by 4π is equal to 9. Take the positive square root of 9 to get $r = 3$. This answer is checking.

PRACTICE 2

1. What is the surface area of a cylinder with a height of 5 m and a base radius of 4 m? (Leave your answer in terms of π.)

2. What is the surface area of a cylinder with a base area of 25π in. and a height of 10 in.? (Leave your answer in terms of π.)

3. What is the surface area of a sphere with a radius of 8 cm? (Leave your answer in terms of π.)

4. What is the measure of the radius of a sphere with a total surface area of 16π cm?

ANSWERS

PRACTICE 1

1. *Read and understand the question.* This question is looking for the surface area of a rectangular prism. Each of the three dimensions is known.
Make a plan. Use the formula $SA = 2lw + 2lh + 2wh$, and substitute the given values for the length, width, and height.
Carry out the plan. The formula becomes $SA = 2(15)(11) + 2(15)(12) + 2(11)(12)$. Multiply to get $SA = 330 + 360 + 264$. Add to get a surface area of 954 cm².
Check your answer. Substitute the values into the formula again to double-check your solution. The formula
$$SA = 2(15)(11) + 2(15)(12) + 2(11)(12)$$
simplifies to
$$2(165) + 2(180) + 2(132) = 330 + 360 + 264 = 954$$
This answer is checking.

2. *Read and understand the question.* This question is looking for the surface area of a rectangular prism. Each of the three dimensions is known.
Make a plan. Use the formula $SA = 2lw + 2lh + 2wh$, and substitute the given values for the length, width, and height.
Carry out the plan. The formula becomes $SA = 2(20)(4) + 2(20)(21) + 2(4)(21)$. Multiply to get $SA = 160 + 840 + 168$. Add to get a surface area of 1,168 m².
Check your answer. To check this answer, substitute the values into the formula again to double-check your solution. The formula
$$SA = 2(20)(4) + 2(20)(21) + 2(4)(21)$$
simplifies to
$$2(80) + 2(420) + 2(84) = 160 + 840 + 168 = 1,168$$
This answer is checking.

3. *Read and understand the question.* This question is looking for the surface area of a cube when the measure of the edge of the cube is given.
   Make a plan. Use the formula $SA = 6e^2$, and substitute the value of e. Remember that the surface area will be represented in square units.
   Carry out the plan. Substitute into the formula to get $SA = 6(9)^2$. Evaluate the exponent first to get $SA = 6(81)$. Multiply to find the surface area: $SA = 486$ in.^{2}.
   Check your answer. To check your answer, divide the total surface area by 6.
   $$486 \div 6 = 81$$
   Then, take the positive square root of 81 to find the measure of the edge of the cube. The positive square root of 81 is 9, so this result is checking.

4. *Read and understand the question.* This question is looking for the measure of the edge of a cube when the total surface area is known.
   Make a plan. Use the formula for the surface area of a cube and work backward to solve for e.
   Carry out the plan. Substitute the values into the formula to get $96 = 6e^2$. Divide each side of the equation by 6 to get $16 = e^2$. Take the positive square root of each side of the equation to find the value of e: $e = 4$ m.
   Check your answer. To check this answer, substitute the value of e into the surface area formula. The formula becomes $SA = 6(4)^2$. Evaluate the exponent to get $SA = 6(16)$. Multiply to get a surface area of 96 m^2. This result is checking.

PRACTICE 2

1. *Read and understand the question.* This question asks for the surface area of a cylinder when the dimensions are given.
   Make a plan. Use the formula for the surface area of a cylinder and substitute the known values. Evaluate to find the surface area.
   Carry out the plan. Because the radius is 4 m, the diameter is 2×4, or 8 m. The formula $SA = 2\pi r^2 + \pi dh$ becomes $SA = 2\pi(4)^2 + \pi(8)(5)$. Evaluate the exponents and multiply to get $SA = 32\pi + 40\pi$. Combine like terms to get $SA = 72\pi$ m^2.
   Check your answer. Substitute again into the formula to double-check your solution. The formula is
   $$SA = 2\pi(4)^2 + \pi(8)(5) = 2\pi(16) + 40\pi = 32\pi + 40\pi = 72\pi.$$
   This answer is checking.

2. *Read and understand the question.* This question asks for the surface area of a cylinder when the area of the bases and the height of the cylinder are given.

Make a plan. Use the formula for the surface area of a cylinder, and substitute the known values. Find the radius and the diameter by using the given area of the bases. Evaluate to find the surface area.

Carry out the plan. Because the area of the bases is 25π, substitute this value for πr^2 in the formula. The radius is equal to the positive square root of 25, so $r = 5$ and $d = 10$. The formula $SA = 2\pi r^2 + \pi dh$ becomes $SA = 2(25\pi) + \pi(10)(10)$. Evaluate to get $SA = 50\pi + 100\pi$. Combine like terms to get $SA = 150\pi$ in.^{2}.

Check your answer. Substitute again into the formula to double-check your solution. The formula is

$$SA = 2\pi(5)^2 + \pi(10)(10) = 2\pi(25) + 100\pi = 50\pi + 100\pi = 150\pi$$

This answer is checking.

3. *Read and understand the question.* This question is looking for the surface area of a sphere when the radius of the sphere is known.

Make a plan. Use the formula $SA = 4\pi r^2$ and substitute $r = 8$.

Carry out the plan. The formula becomes $SA = 4\pi(8)^2$. Evaluate the exponent to get $SA = 4\pi(64)$. Multiply to get the total surface area: $SA = 256\pi$ cm^2.

Check your answer. To check this solution, work backward and divide the total surface area by 4π: 256π divided by 4π is equal to 64. Take the positive square root of 64 to get $r = 8$. This answer is checking.

4. *Read and understand the question.* This question is looking for the radius of the sphere when the surface area is known.

Make a plan. Use the formula $SA = 4\pi r^2$, and substitute the known surface area. Use the strategy of working backward to find the length of the radius, r.

Carry out the plan. The formula becomes $16\pi = 4\pi r^2$. Divide each side of the equation by 4π. The equation becomes $4 = r^2$. Take the positive square root of each side to get $r = 2$. The radius is 2 cm.

Check your answer. To check this solution, substitute the value of r into the surface area formula. The formula becomes

$$SA = 4\pi(2)^2 = 4(4)\,\pi = 16\pi.$$

This answer is checking.

volume word problems

. . . treat Nature by the sphere, the cylinder and the cone . . .
—Paul Cézanne (1839–1906)

This lesson will explain each of the volume formulas for common geometric solids, and how to apply these formulas to word problems with these figures.

VOLUME

The volume of a three-dimensional figure is the amount of space in the figure.

Many volume formulas can be summarized as the area of the base multiplied by the height, or $V = Bh$.

....

TIP: The volume of any figure is expressed in **cubic units**, or **units³**.

....

Rectangular Prism

A rectangular prism is a three-dimensional solid whose faces are all rectangles. The formula for the volume of a rectangular solid is *Volume = length × width × height*, or $V = l \times w \times h$.

Example

What is the volume of a rectangular prism with a height of 10 m, a width of 14 m, and a height of 20 m?

Read and understand the question. This question is looking for the volume of a rectangular prism when the three dimensions are given.

Make a plan. Use the formula $V = l \times w \times h$, and substitute the given values. Then, evaluate the formula.

Carry out the plan. The volume is $V = 10 \times 14 \times 20 = 2{,}800 \ m^3$.

Check your answer. To check this solution, divide the volume by two of the dimensions to check if the result is the third dimension: $2{,}800 \div 10 = 280$; $280 \div 14 = 20$, which is the remaining dimension. This answer is checking.

Cube

A cube is a special type of rectangular prism where each face is the same shape and size. A cube has faces that are all congruent squares, so each edge is the same length. The volume of a cube can be found by using the formula for rectangular prisms, but it can also be found by using the formula $V = e^3$, where e is the measure of an edge of the cube.

Example

The measure of the edge of a cube is 7 cm. What is the volume of the cube?

Read and understand the question. This question is looking for the volume of a cube when the measure of an edge of the cube is known.

Make a plan. Use the formula $V = e^3$, and substitute the given value of e. Then, evaluate the formula.

Carry out the plan. The volume formula becomes $V = (7)^3 = 7 \times 7 \times 7 = 343 \ cm^3$.

Check your answer. To check this solution, divide the volume by the measure of the edge of the cube two times to see if the result is also the measure of the edge:

$343 \div 7 = 49$

$49 \div 7 = 7$

which is the measure of the edge of the cube. This answer is checking.

Triangular Prism

A triangular prism is a three-dimensional solid with triangles as the bases and rectangles as the lateral faces. An example of a triangular prism is shown next.

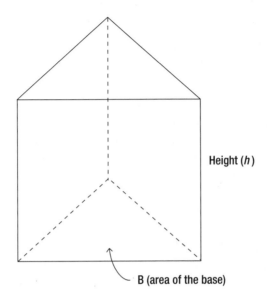

Height (*h*)

B (area of the base)

The formula for the volume of a triangular prism can be found by using the formula $V = Bh$, where B is the area of the base and h is the height of the prism.

Example
What is the volume of a triangular prism with a base area of 12 m² and a height of 4 m?

Read and understand the question. This question is asking for the volume of a triangular prism when the area of each base and the height of the prism are given.

Make a plan. Use the formula $V = Bh$, where B is the area of the base and h is the height of the prism. Substitute the known values and evaluate the formula.

Carry out the plan. The formula becomes $V = 12 \times 4 = 48$. The volume is 48 m³.

Check your answer. To check this problem, divide the volume of the prism by the height, and check to see if the result is the area of the base: $48 \div 4 = 12$, which is the area of the base. This answer is checking.

Pyramid

A pyramid is a geometric solid with a polygon as a base and triangles as each of the other faces. A square base pyramid is shown in the following figure.

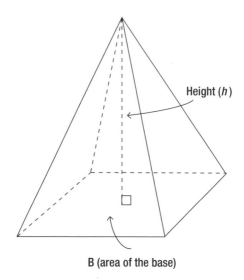

Height (*h*)

B (area of the base)

The volume of a pyramid is equal to one-third of the volume of a prism with the same dimensions. Therefore, the formula for the volume of a pyramid is equal to $V = \frac{1}{3}Bh$, where B is the area of the base and h is the height of the pyramid.

Example
What is the volume of a square base pyramid with a base area of 25 cm^2 and a height of 15 cm?

Read and understand the question. This question is looking for the area of a square base pyramid when the area of the base and the height of the pyramid are known.

Make a plan. Use the formula $V = \frac{1}{3}Bh$, where B is the area of the base and h is the height of the pyramid. Substitute the values and evaluate the formula.

Carry out the plan. The formula becomes

$$V = \frac{1}{3}(25)(15)$$
$$= \frac{1}{3}(375) = 125$$

The volume is 125 cm^3.

Check your answer. To check your answer, work backward. Multiply the volume by 3, and then divide by one of the dimensions to check to see if the result is the other dimension.

$$125 \times 3 = 375$$
$$375 \div 15 = 25$$

This result is checking.

PRACTICE 1

1. What is the volume of a rectangular prism with a length of 5 m, a width of 6 m, and a height of 12 m?

2. What is the height of a rectangular prism with a length of 10 in., a width of 3 in., and a volume of 210 in.^{2}?

3. What is the volume of a cube with an edge that measures 4 cm?

4. What is the volume of a triangular prism with a height of 6 ft. and a base area of 14 ft.^{2}?

5. What is the volume of a square base pyramid with a base area of 18 m^2 and a height of 8 m?

SOLIDS WITH CURVED SURFACES

The next two solids have curved surfaces, and each has one or two bases that are circles.

TIP: Remember that the area of a circle is $A = \pi r^2$. This formula is used to find the volume of the next two solids with bases in the shape of circles.

Cylinder

As mentioned in the previous lesson, a cylinder is a solid with two circles as the bases. To find the volume of a cylinder, multiply the area of the base by the height of the cylinder. Because the base is a circle, the formula is $V = \pi r^2 h$, where r is the radius of the base and h is the height of the cylinder.

Example

The radius of the base of a cylinder is 6 m and the height of the cylinder is 9 m. What is the volume of the cylinder in terms of π?

Read and understand the question. This question is looking for the volume of a cylinder when the radius of the base and the height are known.

Make a plan. Substitute into the volume formula, and then evaluate to find the volume.

Carry out the plan. The formula becomes $V = \pi(6)^2(9)$. Evaluate the exponent to get $V = 36(9)\pi$. Multiply to simplify. The volume is 324π m^3.

Check your answer. To check this solution, use the strategy of working backward and divide the volume by the height times π. Then, take the positive square root to see if this result is the radius of the base: $324\pi \div 9\pi = 36$. The positive square root of 36 is 6, which is the radius of the base. This answer is checking.

Cone

A cone is a solid with one base that is a circle. An example of a cone is shown in the following figure.

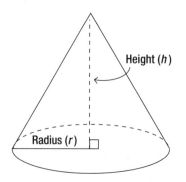

The volume of a cone is equal to one-third of the volume of a cylinder with the same height and the same size circular base. Therefore, the formula for the volume of a cone is $V = \frac{1}{3}\pi r^2 h$.

Example

What is the volume of a cone with a base radius of 9 in. and a height of 5 in.?

Read and understand the question. This question is asking for the volume of a cone when the radius of the base and the height are given.

Make a plan. Use the formula $V = \frac{1}{3}\pi r^2 h$ and substitute the given values. Evaluate the formula to find the volume.

Carry out the plan. The formula becomes $V = \frac{1}{3}\pi(9)^2(5)$. Evaluate the exponent to simplify to $V = \frac{1}{3}(81)(5)\pi$. Multiply. $V = \frac{1}{3}(405)\pi$. Divide 405 by 3 to simplify: $V = 135\pi$ in.^{3}.

Check your answer. To check this answer, work backward by multiplying the volume by 3. Then, divide by the height times π and take the positive square root of the result to get the length of the base radius:

$$135\pi \times 3 = 405\pi$$
$$405\pi \div 5\pi = 81$$

The positive square root of 81 is 9, so this answer is checking.

SPHERE

The volume of a sphere can be found by using the formula $V = \frac{4}{3}\pi r^3$, where r is the radius of the sphere.

Example

What is the volume of a sphere with a radius of 6 in.?

Read and understand the question. This question is looking for the volume of a sphere when the radius is given.

Make a plan. Substitute the value of r into the formula and evaluate to find the volume.

Carry out the plan. The formula becomes $V = \frac{4}{3}\pi(6)^3$. Evaluate the exponent to get $V = \frac{4}{3}\pi(216)$. Multiply to get $V = 288\pi$ in.^{3}.

Check your answer. To check this result, divide the volume by $\frac{4}{3}\pi$. Then, see if the result is the same as 6^3.

$$288\pi \div \frac{4}{3}\pi = 216$$

6^3 is also equal to 216, so this result is checking.

PRACTICE 2

1. What is the volume of a cylinder with a base radius of 2 in. and a height of 4 in.?

2. What is the volume of a cone with a base radius of 5 m and a height of 6 m?

3. What is the volume of a sphere with a radius of 9 cm?

ANSWERS

PRACTICE 1

1. *Read and understand the question.* This question is looking for the volume of a rectangular prism when the three dimensions are given.
   Make a plan. Use the formula $V = l \times w \times h$ and substitute the given values. Then, evaluate the formula.
   Carry out the plan. The volume is $V = 5 \times 6 \times 12 = 360$ m³.
   Check your answer. To check this solution, divide the volume by two of the dimensions to check if the result is the third dimension.
   $$360 \div 5 = 72$$
   $$72 \div 6 = 12$$
   which is the remaining dimension. This solution is checking.
2. *Read and understand the question.* This question is looking for the height of a rectangular prism when the volume and two dimensions are given.
   Make a plan. Use the formula $V = l \times w \times h$ and substitute the given values. Then, evaluate the formula.
   Carry out the plan. The formula becomes $210 = 10 \times 3 \times h$. Multiply to get $210 = 30h$. Divide each side of the equation by 30 to get $h = 7$. The height is 7 in.
   Check your answer. To check this solution, multiply the dimensions to check if the result is the volume.
   $$V = 10 \times 3 \times 7 = 210$$
   This solution is checking.

3. *Read and understand the question.* This question is looking for the volume of a cube when the measure of an edge of the cube is known.

   Make a plan. Use the formula $V = e^3$ and substitute the given value of e. Then, evaluate the formula.

   Carry out the plan. The volume formula becomes $V = (4)^3 = 4 \times 4 \times 4 = 64$ cm^3.

   Check your answer. To check this solution, divide the volume by the measure of the edge of the cube two times to see if the result is also the measure of the edge.

   $$64 \div 4 = 16$$
   $$16 \div 4 = 4$$

   which is the measure of the edge of the cube. This solution is checking.

4. *Read and understand the question.* This question is asking for the volume of a triangular prism when the area of the base and the height of the prism are given.

   Make a plan. Use the formula $V = Bh$, where B is the area of the base and h is the height of the prism. Substitute the known values and evaluate the formula.

   Carry out the plan. The formula becomes $V = 14 \times 6 = 84$. The volume is 84 ft.^{3}.

   Check your answer. To check this problem, divide the volume of the prism by the height and check to see if the result is the area of the base: $84 \div 6 = 14$, which is the area of the base. This answer is checking.

5. *Read and understand the question.* This question is looking for the volume of a square base pyramid when the area of the base and the height of the pyramid are known.

   Make a plan. Use the formula $V = \frac{1}{3}Bh$, where B is the area of the base and h is the height of the pyramid. Substitute the values and evaluate the formula.

   Carry out the plan. The formula becomes
   $$V = \tfrac{1}{3}(18)(8) = \tfrac{1}{3}(144) = 48$$
   The volume is 48 cm^3.

   Check your answer. To check your answer, work backward. Multiply the volume by 3, and then divide by one of the dimensions to check to see if the result is the other dimension:

   $$48 \times 3 = 144$$
   $$144 \div 18 = 8$$

   This result is checking.

PRACTICE 2

1. *Read and understand the question.* This question is looking for the volume of a cylinder when the radius of the base and the height are known.
   Make a plan. Substitute into the volume formula, and then evaluate to find the volume.
   Carry out the plan. The formula becomes $V = \pi(2)^2(4)$. Evaluate the exponent to get $V = 4(4)\pi$. Multiply to simplify. The volume is 16π in.^{3}.
   Check your answer. To check this solution, use the strategy of working backward and divide the volume by the height times π. Then, take the positive square root to see if this result is the radius of the base. $16\pi \div 4\pi = 4$. The positive square root of 4 is 2, which is the radius of the base. This answer is checking.

2. Read and understand the question. This question is asking for the volume of a cone when the radius of the base and the height are given.
   Make a plan. Use the formula $V = \frac{1}{3}\pi r^2 h$ and substitute the given values. Evaluate the formula to find the volume.
   Carry out the plan. The formula becomes $V = \frac{1}{3}\pi(5)^2(6)$. Evaluate the exponent to simplify to $V = \frac{1}{3}(25)(6)\pi$. Multiply: $V = \frac{1}{3}(150)\pi$. Divide 150 by 3 to simplify: $V = 50\pi$ m^3.
   Check your answer. To check this answer, work backward by multiplying the volume by 3. Then, divide by the height times π and take the positive square root of the result to get the length of the base radius.
   $$50\pi \times 3 = 150\pi$$
   $$150\pi \div 6\pi = 25$$
   The positive square root of 25 is 5, so this question is checking.

3. *Read and understand the question.* This question is looking for the volume of a sphere when the radius is given.
   Make a plan. Substitute the value of r into the formula and evaluate to find the volume.
   Carry out the plan. The formula becomes $V = \frac{4}{3}\pi(9)^3$. Evaluate the exponent to get $V = \frac{4}{3}\pi(729)$. Multiply to get $V = 972\pi$ cm^3.
   Check your answer. To check this result, divide the volume by $\frac{4}{3}\pi$. Then, see if the result is the same as 9^3.
   $$972\pi \div \frac{4}{3}\pi = 729$$
   9^3 is also equal to 729, so this result is checking.

coordinate geometry
word problems

Inspiration is needed in geometry, just as much as in poetry.
—ALEKSANDR PUSHKIN (1799–1837)

This lesson will cover the coordinate plane and important topics related to the coordinate plane, such as quadrants, slope, distance, and midpoint.

THE COORDINATE PLANE

The coordinate plane is made up of two perpendicular number lines that intersect, or cross, at a point known as the origin. Every point in the coordinate plane has a location determined by its coordinates, (x,y), and the origin has coordinates $(0,0)$. When these lines intersect, they create four sections called quadrants. The quadrants are numbered, usually with Roman numerals, starting with the upper right-hand quadrant.

··

TIP: The coordinates of points determine in which quadrant they are located. The first value in the parentheses is the x-coordinate and determines the number of units to move right (if a positive number) or left (if a negative number). The second value in the parentheses is the *y*-coordinate and determines the number of units to move up (if a positive number) or down (if a negative number).

....

An example of the coordinate plane with the origin and quadrants labeled is shown below.

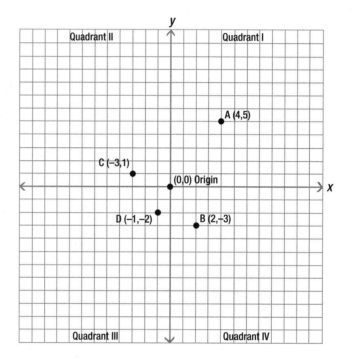

Each point in the plane has a location given by two values in parentheses. To graph a point in the plane, start at the origin. Look at the first value. Move to the right that number of spaces if the first number is positive or to the left if the first number is negative. From there, look at the second value in parentheses. If the number is positive, move up that many spaces or down that many spaces if the number is negative.

For example, to graph the point $A(4,5)$, start at the origin and move four units to the right and five units up. To graph the point $B(2,-3)$, start at the origin and move two spaces to the right and three units down. To graph the point $C(-3,1)$, move three units to the left and one unit up from the origin. To graph the point $D(-1,-2)$, count one unit to the left and two units down from the origin. Each of these points is graphed in the figure.

Use the practice set next to test your skill graphing points and naming quadrants.

PRACTICE 1

Name the quadrant in which each point is located:

(–5,–7)   _____

(4,–2)   _____

(–8,3)   _____

(6,6)   _____

TIP: The formulas in coordinate geometry often use the values (x_1,y_1) and (x_2,y_2) . Use the subscripts with x and y to keep track of each point.

SLOPE FORMULA

The slope of a line can be described as the steepness of the line. Lines that tilt up to the right have positive slope. Lines that tilt up to the left have negative slope. Horizontal lines have zero slope, and vertical lines have undefined slope. To find the slope of a line, first take any two points on the line. Then, to find the slope between the two points, use the formula

$$m = \frac{y_1 - y_2}{x_1 - x_2}$$

Substitute the coordinates of the points into the formula and evaluate.

Example
What is the slope between the points (4,2) and (–3,7)?

Read and understand the question. This question is looking for the slope between two points.
Make a plan. Use the formula

$$m = \frac{y_1 - y_2}{x_1 - x_2}$$

and substitute the coordinates of the two points. Use (4,2) as (x_1,y_1) and (–3,7) as (x_2,y_2).

Carry out the plan. The formula becomes

$$m = \frac{2-7}{4-(-3)} = \frac{-5}{4+3} = \frac{-5}{7}$$

The slope is $-\frac{5}{7}$.

Check your answer. To check this result, substitute the values in the opposite order to make sure that the slope is the same. The formula is

$$m = \frac{7-2}{-3-4} = \frac{5}{-7} = -\frac{5}{7}$$

This result is checking.

DISTANCE FORMULA

Another common formula used in coordinate geometry is the distance formula. To find the distance between any two points, use the formula $d = \sqrt{(x_1 - x_2)^2 + (y_1 - y_2)^2}$.

Example
What is the distance between the two points (1,6) and (–2,2)?

Read and understand the question. This question is looking for the distance between two points.

Make a plan. Use the formula $d = \sqrt{(x_1 - x_2)^2 + (y_1 - y_2)^2}$ and substitute the coordinates of the two points. Use (1,6) as (x_1, y_1) and (–2,2) as (x_2, y_2).

Carry out the plan. The formula becomes

$$d = \sqrt{(1-(-2))^2 + (6-2)^2} = \sqrt{(1+2)^2 + 4^2} = \sqrt{3^2 + 4^2}$$

Evaluate the exponents and add the squares together.

$$\sqrt{9+16} = \sqrt{25}$$

Take the square root of 25 to get a distance of 5 units.

Check your answer. To check this result, substitute the values in the opposite order to make sure that the distance is the same. The formula is

$$d = \sqrt{(-2-1)^2 + (2-6)^2} = \sqrt{(-3)^2 + (-4)^2} = \sqrt{9+16} = \sqrt{25} = 5 \text{ units}$$

This result is checking.

MIDPOINT FORMULA

The midpoint of two points is the location halfway between the points. To find the midpoint between any two points, use the formula $\left(\frac{x_1 + x_2}{2}, \frac{y_1 + y_2}{2}\right)$.

Example
What is the midpoint between the two points (9,3) and (–5,1)?

Read and understand the question. This question is looking for the midpoint between two points.

Make a plan. Use the formula $\left(\frac{x_1 + x_2}{2}, \frac{y_1 + y_2}{2}\right)$ and substitute the coordinates of the two points. Use (9,3) as (x_1, y_1) and (–5,1) as (x_2, y_2).

Carry out the plan. The formula becomes $\left(\frac{9 + -5}{2}, \frac{3 + 1}{2}\right)$. Add to get $\left(\frac{4}{2}, \frac{4}{2}\right)$. Simplify each fraction to get the midpoint of (2,2).

Check your answer. To check this result, substitute the values in the opposite order to make sure that the midpoint is the same. The formula is

$$\left(\frac{-5 + 9}{2}, \frac{1 + 3}{2}\right) = \left(\frac{4}{2}, \frac{4}{2}\right) = (2,2)$$

This result is checking.

PRACTICE 2

1. What is the slope between the points (2,3) and (–3,4)?

2. What is the slope between the points (–3,6) and (–1,–4)?

3. What is the distance between the points (1,2) and (–5,10)?

4. What is the distance between the points (–8,16) and (–3,4)?

5. What is the midpoint between the points (3,5) and (1,3)?

6. What is the midpoint between the points (–4,–2) and (0,6)?

ANSWERS

Practice 1

(−5,−7)	_____III_____
(4,−2)	_____IV_____
(−8,3)	_____II_____
(6,6)	_____I_____

PRACTICE 2

1. *Read and understand the question.* This question is looking for the slope between two points.
   Make a plan. Use the formula $m = \frac{y_1 - y_2}{x_1 - x_2}$ and substitute the coordinates of the two points. Use (2,3) as (x_1,y_1) and (−3,4) as (x_2,y_2).
   Carry out the plan. The formula becomes
   $$m = \frac{3-4}{2-(-3)} = \frac{-1}{2+3} = \frac{-1}{5}$$
   The slope is $-\frac{1}{5}$.
   Check your answer. To check this result, substitute the values in the opposite order to make sure that the slope is the same. The formula is
   $$m = \frac{4-3}{-3-2} = \frac{1}{-5} = -\frac{1}{5}$$
   This result is checking.

2. *Read and understand the question.* This question is looking for the slope between two points.
   Make a plan. Use the formula
   $$m = \frac{y_1 - y_2}{x_1 - x_2}$$
   and substitute the coordinates of the two points. Use (−3,6) as (x_1,y_1) and (−1,−4) as (x_2,y_2).
   Carry out the plan. The formula becomes
   $$m = \frac{6-(-4)}{-3-(-1)} = \frac{6+4}{-3+1} = \frac{10}{-2} = -5$$
   The slope is −5.
   Check your answer. To check this result, substitute the values in the opposite order to make sure that the slope is the same. The formula is
   $$m = \frac{-4-6}{-1-(-3)} = \frac{-10}{-1+3} = \frac{-10}{2} = -5$$
   This result is checking.

3. *Read and understand the question.* This question is looking for the distance between two points.
   Make a plan. Use the formula
   $$d = \sqrt{(x_1 - x_2)^2 + (y_1 - y_2)^2}$$
   and substitute the coordinates of the two points. Use (1,2) as (x_1, y_1) and (–5,10) as (x_2, y_2).
   Carry out the plan. The formula becomes
   $$d = \sqrt{(1-(-5))^2 + (2-10)^2} = \sqrt{(1+5)^2 + (-8)^2} = \sqrt{6^2 + (-8)^2}$$
   Evaluate the exponents and add the squares together: $\sqrt{36+64} = \sqrt{100}$.
   Take the square root of 100 to get a distance of 10 units.
   Check your answer. To check this result, substitute the values in the opposite order to make sure that the distance is the same. The formula is
   $$d = \sqrt{(-5-1)^2 + (10-2)^2} = \sqrt{(-6)^2 + (8)^2} = \sqrt{36+64} = \sqrt{100} = 10 \text{ units}$$
   This result is checking.

4. *Read and understand the question.* This question is looking for the distance between two points.
   Make a plan. Use the formula
   $$d = \sqrt{(x_1 - x_2)^2 + (y_1 - y_2)^2}$$
   and substitute the coordinates of the two points. Use (–8,16) as (x_1, y_1) and (–3,4) as (x_2, y_2).
   Carry out the plan. The formula becomes
   $$d = \sqrt{(-8-(-3))^2 + (16-4)^4} = \sqrt{(-8+3)^2 + 12} = \sqrt{(-5)^2 + 12^2}$$
   Evaluate the exponents and add the squares together: $\sqrt{25+144} = \sqrt{169}$.
   Take the square root of 169 to get a distance of 13 units.
   Check your answer. To check this result, substitute the values in the opposite order to make sure that the distance is the same. The formula is
   $$d = \sqrt{(-3-(-8))^2 + (4-16)^2} = \sqrt{(-3+8)^2 + (-12)^2} = \sqrt{5^2 + (-12)^2}$$
   $$= \sqrt{25+144} = \sqrt{169} = 13 \text{ units}$$
   This result is checking.

5. Read and understand the question. This question is looking for the midpoint between two points.
Make a plan. Use the formula $\left(\frac{x_1 + x_2}{2}, \frac{y_1 + y_2}{2}\right)$ and substitute the coordinates of the two points. Use (3,5) as (x_1,y_1) and (1,3) as (x_2,y_2).

Carry out the plan. The formula becomes $\left(\frac{1+3}{2}, \frac{3+5}{2}\right)$. Add to get $\left(\frac{4}{2}, \frac{8}{2}\right)$. Simplify each fraction to get the midpoint of (2,4).

Check your answer. To check this result, substitute the values in the opposite order to make sure that the midpoint is the same. The formula is

$$\left(\frac{1+3}{2}, \frac{3+5}{2}\right) = \left(\frac{4}{2}, \frac{8}{2}\right) = (2,4)$$

This result is checking.

6. *Read and understand the question.* This question is looking for the midpoint between two points.
Make a plan. Use the formula $\left(\frac{x_1 + x_2}{2}, \frac{y_1 + y_2}{2}\right)$ and substitute the coordinates of the two points. Use (–4,–2) as (x_1,y_1) and (0,6) as (x_2,y_2).

Carry out the plan. The formula becomes $\left(\frac{-4+0}{2}, \frac{-2+6}{2}\right)$. Add to get $\left(\frac{-4}{2}, \frac{4}{2}\right)$. Simplify each fraction to get the midpoint of (–2,2).

Check your answer. To check this result, substitute the values in the opposite order to make sure that the midpoint is the same. The formula is

$$\left(\frac{0+-4}{2}, \frac{6+-2}{2}\right) = \left(\frac{-4}{2}, \frac{4}{2}\right) = (-2,2)$$

This result is checking.

ⓢ ⓔ ⓒ ⓣ ⓘ ⓞ ⓝ 5

probability and statistics word problems— increasing your chances

PROBABILITY IS THE STUDY of uncertainty. When you calculate the theoretical probability of an event, you are actually making a guess on the likeliness of the event happening. The meteorologist makes a prediction about how likely it is to rain or snow, or the chance it will be a bright, sunny day. When playing card and board games, you can use strategies to increase your chance of winning, based on these principles of probability.

Statistics, on the other hand, is a little more of an exact science. For example, when you are calculating the mean, or average, of a set of data, you can use the actual data values. When you are in school, you can use statistics in a number of ways. You may find the mean of your test scores to see how well you are doing in a class or the range in the heights of your classmates. A realtor may use the median home prices in a certain area to help a homebuyer determine if a price is fair.

The word problems in this section detail the many types of probability and statistics problems commonly found in our everyday lives, as well as the math classroom. Increase your chances of success by reading through this section and completing the practice problems in each lesson.

This section will cover probability and statistics word problems including:

- basic probability
- independent and dependent events

- counting problems
- permutations
- combinations
- measures of central tendency—mean, median, mode
- range of data sets

ⓛ ⓔ ⓢ ⓢ ⓞ ⓝ 27

basic probability word problems

Life is a school of probability.
—WALTER BAGEHOT (1826–1877)

This lesson will cover the basics of probability and will show you how to apply these concepts to word problems.

SIMPLE PROBABILITY

The probability of a certain event can be found by using the formula:

$$P(E) = \frac{\text{number of ways an event can occur}}{\text{total number of possible outcomes}}$$

To find the total number of possible outcomes, make a list of the possibilities. This list is called a sample space. For example, when you are flipping a coin, the sample space is {heads, tails} and when you are rolling a number cube, the sample space is {1, 2, 3, 4, 5, 6}.

Now, let's practice finding the probability of some simple events. For example, when you are flipping a coin, there are two sides and one of the sides is heads. Therefore, the probability of getting heads when flipping a coin is $\frac{1}{2}$.

When you are rolling a number cube, there are six sides and one of them has a 4 on it. Therefore, the probability of rolling a number cube and getting a 4 is $\frac{1}{6}$.

Use the steps to solving word problems to solve the following simple probability question:

What is the probability of selecting a black card at random from a standard deck of 52 cards?

Read and understand the question. This question is looking for the probability of selecting one black card from a standard deck. There are 52 cards in a standard deck.

Make a plan. Use the fact that there are 52 cards in the deck and the fact that 26 of those cards are black. Substitute into the formula
$$P(E) = \frac{\text{number of ways an event can occur}}{\text{total number of possible outcomes}}.$$
Carry out the plan. The formula becomes $P(\text{black card}) = \frac{26}{52} = \frac{1}{2}$. The probability is equal to $\frac{1}{2}$.

Check your answer. To check this answer, substitute the values again to see if the answer is reasonable. Since 26 out of the 52 cards of the deck are black, half of the cards are black. Thus, the probability is equal to $\frac{1}{2}$. This answer is checking.

FINDING THE PROBABILITY THAT AN EVENT WILL NOT HAPPEN

The probability of an event not happening is always equal to the probability of the event happening subtracted from one. The formula is $P(\text{not } E) = 1 - P(E)$. For example, if the probability of selecting a heart from a standard deck of cards is $\frac{1}{4}$, then the probability of not selecting a heart is $1 - \frac{1}{4} = \frac{3}{4}$.

PRACTICE 1

1. Charles is spinning a spinner like the one shown in the figure while he is playing a game.

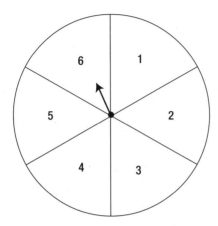

What is the probability that he spins a 5 on his next turn?

2. The letters from the word GEOMETRY are placed in a hat. What is the probability of selecting an *M* when you are choosing one letter without looking?

3. What is the probability of getting a 7 when you are rolling a number cube?

4. If the probability that it will rain is $\frac{1}{3}$, then what is the probability that it will **not** rain?

5. When you are selecting a number at random from the digits 0–9, what is the probability of **not** selecting a 3?

COMPOUND PROBABILITY ("OR" STATEMENTS)

When you are finding the probability of events with two or more conditions, add the probabilities together. For example, to find the probability of rolling a 2 or a 3 when you are rolling a number cube, the probability of rolling a 2 is $\frac{1}{6}$ and the probability of rolling a 3 is $\frac{1}{6}$. Thus, the probability of rolling a 2 or a 3 is $P(2 \text{ or } 3) = \frac{1}{6} + \frac{1}{6} = \frac{2}{6} = \frac{1}{3}$. Because these two events could not occur at the same time, these are known as *mutually exclusive* events.

> **TIP:** In probability questions, the key word *or* tells you to add the probabilities together.

What about events that are not mutually exclusive? Take, for example, the event of rolling a number cube. What is the probability that you roll a 4 or an even number? This time, rolling a 4 satisfies the first condition and rolling a 2, 4, or 6 satisfies the second condition. Because rolling a 4 satisfies both conditions, these events are not mutually exclusive. Therefore, to find the probability of rolling a 4 or an even, count the number of ways each event can happen and subtract the number of ways that both conditions are satisfied. The probability of rolling a 4 or an even number is equal to $\frac{1}{6} + \frac{3}{6} - \frac{1}{6} = \frac{3}{6} = \frac{1}{2}$.

PRACTICE 2

1. What is the probability of selecting a red or a blue marble from a jar with 3 red marbles, 4 green marbles, and 2 blue marbles?

2. What is the probability of selecting a *P* or a vowel from the word PROBABILITY?

3. What is the probability of selecting an *I* or any vowel from the word PROBABILITY?

4. When you are selecting a card at random from a standard deck of 52 cards, what is the probability of selecting a king or a red card?

ANSWERS

PRACTICE 1

1. *Read and understand the question.* This question is looking for the probability of getting a 5 when you are spinning a spinner. There are 6 equal sections to the spinner.
   Make a plan. Use the fact that there are 6 equal sections and 1 of the sections is labeled with a 5. Substitute into the formula
   $$P(E) = \frac{\text{number of ways an event can occur}}{\text{total number of possible outcomes}}.$$
   Carry out the plan. The formula becomes $P(5) = \frac{1}{6}$. The probability is equal to $\frac{1}{6}$.
   Check your answer. To check this answer, substitute the values again to see if the answer is reasonable. Since 1 out of the 6 sections is labeled with a 5, the probability is equal to $\frac{1}{6}$. This answer is checking.

2. *Read and understand the question.* This question is looking for the probability of selecting a letter M from the word GEOMETRY. There are a total of 8 letters in the word.
   Make a plan. Use the facts that there are 8 letters in the word and 1 is a letter M. Substitute into the formula
   $$P(E) = \frac{\text{number of ways an event can occur}}{\text{total number of possible outcomes}}.$$
   Carry out the plan. The formula becomes $P(M) = \frac{1}{8}$. The probability is equal to $\frac{1}{8}$.
   Check your answer. To check this answer, substitute the values again to see if the answer is reasonable. Since 1 out of the 8 letters in the word is the letter M, the probability is equal to $\frac{1}{8}$. This answer is checking.

3. *Read and understand the question.* This question is looking for the probability of rolling a 7 on a number cube. There are 6 sides on a number cube.
   Make a plan. Use the fact that there are 6 sides on the number cube and none of the sides have 7 dots. Substitute into the formula
   $$P(E) = \frac{\text{number of ways an event can occur}}{\text{total number of possible outcomes}}.$$
   Carry out the plan. The formula becomes $P(7) = \frac{0}{6} = 0$. The probability is equal to 0.
   Check your answer. This was an example of an impossible event. Thus, the probability is equal to 0. This answer is checking.

4. *Read and understand the question.* This question is looking for the probability that it will not rain when the probability that it will rain is given.
   Make a plan. Use the fact that the probability that it will rain is $\frac{1}{3}$. Substitute into the formula P(not E) = 1 – P(E).
   Carry out the plan. The formula becomes P(not rain) = $1 - \frac{1}{3} = \frac{3}{3} - \frac{1}{3} = \frac{2}{3}$. The probability is equal to $\frac{2}{3}$.
   Check your answer. To check this answer, add the probability that it will rain to the probability that it will not rain. Make sure that the sum is equal to 1: $\frac{1}{3} + \frac{2}{3} = \frac{3}{3} = 1$. This answer is checking.

5. *Read and understand the question.* This question is looking for the probability of not selecting a 3 from the digits 0 – 9. There are 10 digits from 0 to 9.
   Make a plan. Use the fact that the probability of selecting a 3 is $\frac{1}{10}$. Substitute into the formula P(not E) = 1 – P(E).
   Carry out the plan. The formula becomes P(not 3) = $1 - \frac{1}{10} = \frac{10}{10} - \frac{1}{10} = \frac{9}{10}$. The probability is equal to $\frac{9}{10}$.
   Check your answer. To check this answer, add the probability of selecting a 3 with the probability of not selecting a 3, and make sure that the sum is equal to 1: $\frac{1}{10} + \frac{9}{10} = \frac{10}{10} = 1$. This answer is checking.

PRACTICE 2

1. *Read and understand the question.* This question is looking for the probability of selecting a red or a blue marble when you are picking a marble without looking. There are 3 red marbles, 4 green marbles, and 2 blue marbles in the jar.
   Make a plan. Use the fact that there are a total of 9 marbles in the jar; 3 of them are red and 2 of them are blue. The key word *or* in the question tells you to add the probabilities together, and in this problem a marble cannot be red and blue at the same time, so the events are mutually exclusive. Use the basic formula $P(E) = \frac{\text{number of ways an event can occur}}{\text{total number of possible outcomes}}$ and add the probabilities together.
   Carry out the plan. The formula becomes P(red or blue) = $\frac{3}{9} + \frac{2}{9} = \frac{5}{9}$. The probability is equal to $\frac{5}{9}$.
   Check your answer. To check this answer, substitute the values again to see if the answer is reasonable. Since 5 out of the 9 marbles are either red or blue, this is equal to the probability of $\frac{5}{9}$. This answer is checking.

2. *Read and understand the question.* This question is looking for the probability of selecting a *P* or a vowel from the letters of the word PROBABILITY. There are a total of 11 letters, one *P*, and 4 vowels.

   Make a plan. Use the fact that there are a total of 11 letters in the word; 1 of them is a *P* and 4 of them are vowels. The key word *or* in the question tells you to add the probabilities together. *P* is not also a vowel, so the events are mutually exclusive. Use the basic formula $P(E) = \frac{\text{number of ways an event can occur}}{\text{total number of possible outcomes}}$ and add the probabilities together.

   Carry out the plan. The formula becomes
   $$P(P \text{ or a vowel}) = \tfrac{1}{11} + \tfrac{4}{11} = \tfrac{5}{11}$$
   The probability is equal to $\tfrac{5}{11}$.

   Check your answer. To check this answer, substitute the values again to see if the answer is reasonable. Since 5 out of the 11 letters are either the letter *P* or a vowel, this is equal to the probability of $\tfrac{5}{11}$. This answer is checking.

3. *Read and understand the question.* This question is looking for the probability of selecting an *I* or any vowel from the letters of the word PROBABILITY. There are a total of 11 letters, 2 *Is*, and 4 vowels, including the *Is*.

   Make a plan. Use the fact that there are a total of 11 letters in the word; 2 of them are *Is* and a total of 4 are vowels. The key word *or* in the question tells you to add the probabilities together. *I* is also a vowel, so the events are *not* mutually exclusive. Use the basic formula $P(E) = \frac{\text{number of ways an event can occur}}{\text{total number of possible outcomes}}$ and add the probabilities together, but then subtract the probability of the two *Is* because they are also vowels.

   Carry out the plan. The formula becomes
   $$P(I \text{ or any vowel}) = \tfrac{2}{11} + \tfrac{4}{11} - \tfrac{2}{11} = \tfrac{4}{11}$$
   The probability is equal to $\tfrac{4}{11}$.

   Check your answer. To check this answer, substitute the values again to see if the answer is reasonable. Since 4 out of the 11 letters are vowels including the letter *I*, this is equal the probability of $\tfrac{4}{11}$. This answer is checking.

4. *Read and understand the question.* This question is looking for the probability of selecting a king or a red card when you are choosing from a deck of
52 cards. There are 26 red cards and 4 kings in the deck.
Make a plan. Use the facts that there are a total of 52 cards in the deck
and 26 of them are red and 4 of them are kings. The key word *or* in
the question tells you to add the probabilities together. In this problem,
there are two cards that are red and a king at same time, so the
events are not mutually exclusive. Use the basic formula
$$P(E) = \frac{\text{number of ways an event can occur}}{\text{total number of possible outcomes}}$$ and add the probabilities together, but
subtract the probability of selecting a king that is also a red card.
Carry out the plan. The formula becomes
$$P(\text{red card or king}) = \frac{26}{52} + \frac{4}{52} - \frac{2}{52} = \frac{28}{52} = \frac{7}{13}$$
The probability is equal to $\frac{7}{13}$.
Check your answer. To check this answer, substitute the values again to see
if the answer is reasonable. Since 28 out of the 52 cards are either red or
kings, this is equal to the probability of $\frac{28}{52} = \frac{7}{13}$. This answer is checking.

independent and dependent events

To us probability is the very guide of life.
—Bishop Joseph Butler (1692–1752)

This lesson will explain the difference between independent and dependent events, and provide examples of word problems on these topics.

INDEPENDENT EVENTS

Independent events are a set of events where the probability of the second event does not depend on the outcome of the first. In this situation, the probabilities for each individual event are multiplied together.

....

TIP: The formula to find the probability of two independent events A and B is P(A and B) = P(A) × P(B). The key word *and* tells you to multiply the probability for each event to find the probability that both events will happen.

....

Take the following situation:

A jar contains 4 blue marbles, 3 red marbles, and 1 green marble. One marble is selected, then replaced, and then a second marble is chosen. Find the probability for each situation.

What is the probability that a blue marble is selected, and then a green marble is selected?

Read and understand the question. This question is looking for the probability of selecting a blue marble first, and then a green marble second. The events are independent because the first marble is replaced.

Make a plan. Use the facts that there are 4 blue marbles and 1 green marble, and there are a total of 8 marbles. Find the probability of each event and multiply them.

Carry out the plan. The probability of selecting a blue marble at random is $\frac{4}{8} = \frac{1}{2}$, and the probability of selecting a green marble at random is $\frac{1}{8}$. Multiply these probabilities to find the probability that both events will happen: $\frac{1}{2} \times \frac{1}{8} = \frac{1}{16}$.

Check your answer. To check this solution, multiply the probabilities again. Because $\frac{1}{2} \times \frac{1}{8} = \frac{1}{16}$, this solution is checking.

What is the probability that two blue marbles are selected?

Read and understand the question. This question is looking for the probability of selecting a blue marble first, and then another blue marble second. The events are independent because the first marble is replaced.

Make a plan. Use the facts that there are 4 blue marbles, and there are a total of 8 marbles. Find the probability of each event and multiply them.

Carry out the plan. The probability of selecting a blue marble at random is $\frac{4}{8} = \frac{1}{2}$. Multiply the probability of drawing a blue marble by itself to find the probability that both events will happen: $\frac{1}{2} \times \frac{1}{2} = \frac{1}{4}$.

Check your answer. To check this solution, multiply the probabilities together again. Because $\frac{1}{2} \times \frac{1}{2} = \frac{1}{4}$, this solution is checking.

....

TIP: Problems with replacement are independent events.

....

PRACTICE 1

1. One card is selected at random from a standard deck of 52 cards. The card is replaced, and then another card is selected. What is the probability that the first card chosen is a red card and the second card chosen is a queen?

2. One card is selected at random from a standard deck of 52 cards. The card is replaced, and then another card is selected. What is the probability that the first card chosen is a heart and the second card chosen is a diamond?

3. A red die and a white die are each rolled while playing a board game. What is the probability of getting a 1 on the red die and a 4 on the white die?

4. A game uses 26 letter tiles. Each tile has one letter on it and each letter of the alphabet appears once on the tiles. If one tile is selected at random, replaced, and then another selected, what is the probability that the first tile is an *A* and the second another vowel?

DEPENDENT EVENTS

Dependent events are events where the probability of the event following another depends upon the outcome of the previous event. In other words, the likelihood of an event is determined by what is selected first. These problems are commonly known as dependent events because the second event *depends* on the first.

TIP: The formula to find the probability of two dependent events A and B is also P(A and B) = P(A) × P(B). The key word *and* tells you to multiply the probability for each event together to find the probability that both events happen. However, the probability of the second event will depend upon the outcome of the first event, so be mindful of this fact as you calculate the chances of each event.

Now, let's use the same situation mentioned earlier in this lesson, except this time, the first marble is not replaced.

A jar contains 4 blue marbles, 3 red marbles, and 1 green marble. One marble is selected, not replaced, and then a second marble is chosen. Find the probability for each situation.

What is the probability that a blue marble is selected, and then a green marble is selected?

Read and understand the question. This question is looking for the probability of selecting a blue marble first, and a green marble second. The events are dependent events because the first marble is not replaced after it is selected.

Make a plan. Use the facts that there are 4 blue marbles and 1 green marble, and there are a total of 8 marbles. Because the first marble is not replaced after it is chosen, the number of marbles decreases by one for the second draw. Find the probability of each event and multiply them.

Carry out the plan. The probability of first selecting a blue marble at random is $\frac{4}{8} = \frac{1}{2}$, which leaves only 7 marbles left. The probability of selecting a green marble second is $\frac{1}{7}$. Multiply these probabilities to find the probability that both events will happen: $\frac{1}{2} \times \frac{1}{7} = \frac{1}{14}$.

Check your answer. To check this solution, multiply the probabilities again. Because $\frac{1}{2} \times \frac{1}{7} = \frac{1}{14}$, this answer is checking.

What is the probability that two blue marbles are selected?

Read and understand the question. This question is looking for the probability of selecting a blue marble first, and another blue marble second. The events are dependent events because the first marble is not replaced after it is selected.

Make a plan. Use the facts that there are 4 blue marbles, and there are a total of 8 marbles. Because the first marble is not replaced after it is chosen, the number of marbles decreases by one for the second draw. Find the probability of each event and multiply them together.

Carry out the plan. The probability of first selecting a blue marble at random is $\frac{4}{8} = \frac{1}{2}$, which leaves only 7 marbles. The probability of selecting another blue marble second is $\frac{3}{7}$. Multiply these probabilities to find the probability that both events will happen: $\frac{1}{2} \times \frac{3}{7} = \frac{3}{14}$.

Check your answer. To check this solution, multiply the probabilities together again. Because $\frac{1}{2} \times \frac{3}{7} = \frac{3}{14}$, this solution is checking.

What is the probability that two green marbles are selected?

Read and understand the question. This question is looking for the probability of selecting a green marble first, and another green marble second. The events are dependent events because the first marble is not replaced after it is selected.

Make a plan. Use the facts that there is only 1 green marble, and there are a total of 8 marbles. Because the first marble is not replaced after it is chosen, the number of marbles decreases by one for the second draw. Find the probability of each event and multiply them.

Carry out the plan. The probability of first selecting a green marble at random is $\frac{1}{8}$, which leaves only 7 marbles, none of which are green. Thus, the probability of selecting a green marble second is $\frac{0}{7}$. Multiply these probabilities to find the probability that both events will happen: $\frac{1}{8} \times \frac{0}{7} = \frac{0}{56} = 0$.

Check your answer. To check this solution, use the fact that there is only one green marble. The probability of selecting two green marbles without replacing the first is an impossible event. The probability of any impossible event is equal to 0. This answer is checking.

....

TIP: Problems without replacement are dependent events.

....

PRACTICE 2

1. One card is selected at random from a standard deck of 52 cards. The card is not replaced, and then another card is selected. What is the probability that the first card chosen is a red card and the second card chosen is a queen, assuming that the first card chosen was not a red queen?

2. One card is selected at random from a standard deck of 52 cards. The card is not replaced, and then another card is selected. What is the probability that both cards selected are hearts?

3. A game uses 26 letter tiles. Each tile has one letter on it and each letter of the alphabet appears once on the tiles. If one tile is selected at random, not replaced, and then another selected, what is the probability that the first tile is an A and the second another vowel?

4. At a picnic, a cooler contains 6 cans of apple juice and 4 cans of grape juice. Jerome selects a can of juice, drinks it, and then selects another can. What is the probability that both cans were apple juice?

ANSWERS

PRACTICE 1

1. *Read and understand the question.* This question is looking for the probability of selecting a red card first, and then a queen second. The events are independent events because the first card is replaced after it is selected. *Make a plan.* Use the facts that there are 26 red cards and 4 queens in the deck, and there are a total of 52 cards. Because the first card is replaced after it is chosen, the number of total possible outcomes is always the same (52). Find the probability of each event and multiply them. *Carry out the plan.* The probability of first selecting a red card at random is $\frac{26}{52} = \frac{1}{2}$. The probability of selecting a queen second is $\frac{4}{52} = \frac{1}{13}$. Multiply these probabilities to find the probability that both events will happen: $\frac{1}{2} \times \frac{1}{13} = \frac{1}{26}$. *Check your answer.* To check this solution, multiply the probabilities again. Because $\frac{1}{2} \times \frac{1}{13} = \frac{1}{26}$, this answer is checking.

2. *Read and understand the question.* This question is looking for the probability of selecting a heart first, and a diamond second. The events are independent events because the first card is replaced after it is selected. *Make a plan.* Use the facts that there are 13 hearts and 13 diamonds in the deck, and there are a total of 52 cards. Because the first card is replaced after it is chosen, the number of total possible outcomes is always the same (52). Find the probability of each event and multiply them. *Carry out the plan.* The probability of first selecting a heart at random is $\frac{13}{52} = \frac{1}{4}$. The probability of selecting a diamond second is $\frac{13}{52} = \frac{1}{4}$. Multiply these probabilities to find the probability that both events will happen: $\frac{1}{4} \times \frac{1}{4} = \frac{1}{16}$. *Check your answer.* To check this solution, multiply the probabilities again. Because $\frac{1}{4} \times \frac{1}{4} = \frac{1}{16}$, this solution is checking.

3. *Read and understand the question.* This question is looking for the probability of rolling a 1 on the red die and a 4 on the white die. The events are independent events because the outcome of one die does not depend on the outcome of the other. *Make a plan.* Use the facts that there are 6 sides to a die, and one side has 1 dot and one side has 4 dots. Find the probabilities for each event and multiply them.

Carry out the plan. The probability of rolling a 1 on the red die is $\frac{1}{6}$, and the probability of rolling a 4 on the white die is also $\frac{1}{6}$. Multiply these to find the total probability: $\frac{1}{6} \times \frac{1}{6} = \frac{1}{36}$.

Check your answer. To check this solution, multiply the probabilities again. Because $\frac{1}{6} \times \frac{1}{6} = \frac{1}{36}$, this solution is checking.

4. *Read and understand the question.* This question is looking for the probability of selecting an *A* and another vowel when choosing two letter tiles at random. Because the first tile is replaced, the events are independent events.

Make a plan. Use the fact that there are 26 different tiles, each with a different letter. There is one tile with an *A* and 5 tiles in total with vowels. Find the probability of each situation and multiply them together.

Carry out the plan. The probability of selecting an *A* is $\frac{1}{26}$ and the probability of selecting a vowel is $\frac{5}{26}$. The probability of both events happening is $\frac{1}{26} \times \frac{5}{26} = \frac{5}{676}$.

Check your answer. To check this solution, multiply the probabilities together again. Because $\frac{1}{26} \times \frac{5}{26} = \frac{5}{676}$, this solution is checking.

PRACTICE 2

1. *Read and understand the question.* This question is looking for the probability of selecting a red card first, and a queen second. The events are dependent events because the first card is not replaced after it is selected.

Make a plan. Use the facts: there are 26 red cards and 4 queens in the deck, and there are a total of 52 cards. Because the first card is not replaced after it is chosen, the number of total possible outcomes decreases by one after the first card is selected. Find the probability of each event and multiply them.

Carry out the plan. The probability of first selecting a red card at random is $\frac{26}{52} = \frac{1}{2}$. The probability of selecting a queen second is $\frac{4}{51}$. Multiply these probabilities to find the probability that both events will happen: $\frac{1}{2} \times \frac{4}{51} = \frac{4}{102} = \frac{2}{51}$.

Check your answer. To check this solution, multiply the probabilities again. Because $\frac{1}{2} \times \frac{4}{51} = \frac{4}{102} = \frac{2}{51}$, this solution is checking.

2. *Read and understand the question.* This question is looking for the probability of selecting a heart first, and another heart second. The events are dependent events because the first card is not replaced after it is selected. *Make a plan.* Use the facts that there are 13 hearts in the deck, and there are a total of 52 cards. Because the first card is not replaced after it is chosen, the number of total possible outcomes decreases by one after the first card is selected. In addition, there will be one less heart in the deck. Find the probability of each event and multiply them. *Carry out the plan.* The probability of first selecting a heart at random is $\frac{13}{52} = \frac{1}{4}$. The probability of selecting another heart second is $\frac{12}{51} = \frac{4}{17}$. Multiply these probabilities to find the probability that both events will happen: $\frac{1}{4} \times \frac{4}{17} = \frac{4}{68} = \frac{1}{17}$. *Check your answer.* To check this solution, multiply the probabilities again. Because $\frac{1}{4} \times \frac{4}{17} = \frac{4}{68} = \frac{1}{17}$, this solution is checking.

3. *Read and understand the question.* This question is looking for the probability of selecting an *A* and then another vowel when choosing two letter tiles at random. Because the first tile is not replaced, the events are dependent events.

   Make a plan. Use the fact that there are 26 different tiles, each with a different letter. There is one tile with an *A*, and 5 tiles in total with vowels. If an *A* is selected first and not replaced, then there are only 4 vowels left and a total of 25 tiles. Find the probability of each situation and multiply them. *Carry out the plan.* The probability of selecting an *A* is $\frac{1}{26}$ and the probability of selecting another vowel after an *A* is chosen is $\frac{4}{25}$. The probability of both events happening is $\frac{1}{26} \times \frac{4}{25} = \frac{4}{650} = \frac{2}{325}$. *Check your answer.* To check this solution, multiply the probabilities again. Because $\frac{1}{26} \times \frac{4}{25} = \frac{4}{650} = \frac{2}{325}$, this solution is checking.

4. *Read and understand the question.* This question is looking for the probability that two cans of apple juice are selected at random from a cooler. Because the first can is not replaced, the events are dependent events. *Make a plan.* Use the facts that there are 6 cans of apple juice and 4 cans of grape juice, for a total of 10 cans in the cooler. The first can of juice will not be replaced, so the number of cans of apple juice and the total number of cans will each decrease by one when selecting the second can. *Carry out the plan.* The probability of selecting a can of apple juice for the first drink is $\frac{6}{10} = \frac{3}{5}$. The probability that the second drink is also apple juice is $\frac{5}{9}$. Multiply the probabilities: $\frac{3}{5} \times \frac{5}{9} = \frac{15}{45} = \frac{1}{3}$. *Check your answer.* To check this solution, multiply the probabilities again. Because $\frac{3}{5} \times \frac{5}{9} = \frac{15}{45} = \frac{1}{3}$, this solution is checking.

word problems with the counting principle, permutations, and combinations

*The man ignorant of mathematics will be increasingly limited
in his grasp of the main forces of civilization.*
—JOHN KEMENY (1926–1992)

This lesson will cover the counting principle, permutations, and combinations.
The steps to solving word problems involving these topics will be detailed.

THE COUNTING PRINCIPLE

The counting principle states that the total number of possibilities can be found
by multiplying the number of choices. This was a topic discussed in Lesson 2.

In addition to this strategy, another way to find the total number of possibilities is to take the number of choices in each category and multiply them.

Take the following example.

Sean has three ways to get to school. He can take the bus, get a ride, or
walk. If he can take one of the same three ways to get home, how many total
choices does he have to get from home to school and back?

Read and understand the question. This question is looking for the total number of possibilities when there are three different modes of transportation to school and back.

Make a plan. The trip from home to school is one route he needs to take, and the trip from school back to his home is the second route he needs to take. Take the total number of choices for each route and multiply them.

Carry out the plan. There are three different ways to get from home to school, and three different ways to get from school back to home. Multiply these choices: $3 \times 3 = 9$ different ways.

Check your answer. To check this answer, divide the total number of outcomes by the number of choices in the first route. This becomes $9 \div 3 = 3$, which is the number of choices in the second route. This answer is checking.

PRACTICE 1

1. Charlotte is picking out her class ring. She can select from a ruby, an emerald, or an opal stone, and she can also select silver or gold for the metal. How many different combinations of one stone and one type of metal can she choose?

2. Pete has 5 different ties that match with 3 different shirts. How many shirt-and-tie combinations can he make if he selects one shirt and one tie?

3. How many different sandwiches can be made with 2 choices of bread, 3 choices of toppings, and 2 choices of meat if one choice is selected from each category?

PERMUTATIONS

A permutation of objects is the total number of different orders of the objects. In other words, it is the number of arrangements that can be made when you are working with a certain set of objects.

The number of permutations of a set of objects is based on the counting principle, so take the number of choices for each placement in the set and multiply them. This will give the total number of permutations of the set.

For example, find the total number of permutations (orders) of the set of letters A, C, and T. The permutations do not have to form regular words.

Read and understand the question. This question is looking for the total number of permutations, or orders, of three different letters. Each letter is used exactly once.

Make a plan. Multiply the number of choices for each placement of the letters. A letter can only be used once.

Carry out the plan. There are three letters to choose from for the first letter, two letters to choose from for the second letter, and thus only one letter to choose from for the third letter. The number of permutations is therefore $3 \times 2 \times 1 = 6$.

Check your answer. One way to check this solution is to use the strategy of making an organized list. The possible orders of the letters are ACT, ATC, CAT, CTA, TAC, and TCA. There are 6 different orders, so this solution is checking.

TIP: When multiplying the number of choices to find the permutations of the objects, the number of choices always starts with the total number of objects and decreases by one each time. For example, when you are finding the number of permutations of 5 objects taken 5 at a time, multiply $5 \times 4 \times 3 \times 2 \times 1 = 120$ different permutations.

However, sometimes not all of the objects are used in each permutation. For instance, when you are finding the number of permutations of 5 objects taken only 3 at a time, start with 5 choices and multiply just the first 3 values together. The number of permutations in this situation is $5 \times 4 \times 3 = 60$.

PRACTICE 2

1. How many different permutations are there of 4 objects taken 4 at a time?

2. There are 6 trophies on a shelf. How many different orders of all 6 trophies can be made?

3. How many ways can 7 students come in first, second, and third place in a geography contest, if only 1 student can earn each place?

COMBINATIONS

The major difference between the combinations of a set of objects and the permutations of the same set of objects is that with combinations, the order does not matter. When the order of the objects changes, it is still considered the same combination of items.

Calculating the total number of combinations is very similar to finding the number of permutations. However, because the order does not matter, you must divide by the total number of different orders of the objects.

....

TIP: Because the order does not matter, the total number of combinations of n objects taken n at a time is always equal to 1. For example, the total number of combinations of 3 objects when all 3 objects are used is 1. Even though you can list the objects in different orders, the set still contains the same 3 objects.

....

Take the next example.

How many different combinations of 3 students can be selected for a committee when there are a total of 10 students from which to choose?

Read and understand the question. This question is looking for the total number of combinations of 10 students selected 3 at a time for a committee.

Make a plan. Because the order is not important, find the number of permutations of 10 students taken 3 at a time, and then divide by the number of permutations of 3 students taken 3 at a time.

Carry out the plan. The number of permutations of 10 students taken 3 at a time is equal to $10 \times 9 \times 8$ and the total number of permutations of 3 students taken 3 at a time is equal to $3 \times 2 \times 1$. Divide the number of permutations of 10 students taken 3 at a time by the number of permutations of 3 students taken 3 at a time to find the total number of combinations of 10 students taken 3 at a time.

$$\frac{10 \times 9 \times 8}{3 \times 2 \times 1} = \frac{720}{6} = 120$$

different combinations.

Check your answer. To check this solution, find the number of combinations of 10 students taken 7 at a time, which should be the same result. To do this, calculate the number of permutations of 10 students taken 7 at a time and divide this amount by the permuations of 7 students taken 7 at a time. The number of combinations is equal to

$$\frac{10 \times 9 \times 8 \times 7 \times 6 \times 5 \times 4}{7 \times 6 \times 5 \times 4 \times 3 \times 2 \times 1} = \frac{604,800}{5,040} = 120$$

so this solution is checking.

> **TIP:** When you are looking at the total number of **permutations** of a set of objects, the order matters. When looking at the total number of **combinations** of a set of objects, the order does not matter.

PRACTICE 3

1. How many ways can 2 books be selected out of a series of 4 books if the order is not important?

2. Tyler and his family are going on a trip. He would like to select some movies to watch in the car while traveling. How many different combinations of movies are there if he selects 4 movies to watch out of a total of 8?

3. There are 6 students in a club. For a certain activity, 4 of the students need to form a separate group. How many ways can this group of 4 students be formed?

ANSWERS

PRACTICE 1

1. *Read and understand the question.* This question is looking for the total number of possibilities when there are 3 different choices of stone and 2 choices of metal for a ring.
   Make a plan. Take the total number of choices for each category and multiply them.
   Carry out the plan. There are 3 different choices of stone, and two different choices of metal. Multiply these choices: $3 \times 2 = 6$ different ways to design the ring.

Check your answer. To check this answer, divide the total number of outcomes by the number of choices in the first category. This becomes 6 ÷ 3 = 2, which is the number of choices of metal. This answer is checking.

2. *Read and understand the question.* This question is looking for the total number of possibilities when there are 5 different choices of ties and 3 different choices of shirts.

Make a plan. Take the total number of choices for each category and multiply them.

Carry out the plan. There are 5 different choices of ties, and 3 different choices of shirts. Multiply these choices: $5 \times 3 = 15$ different shirt-and-tie combinations.

Check your answer. To check this answer, divide the total number of outcomes by the number of choices for a tie. This becomes 15 ÷ 5 = 3, which is the number of choices of a shirt. This answer is checking.

3. *Read and understand the question.* This question is looking for the total number of possibilities when making a sandwich from 2 choices of bread, 3 choices of toppings, and 2 choices of meat. One selection from each category will be chosen.

Make a plan. Take the total number of choices for each category and multiply them.

Carry out the plan. There are 2 choices of bread, 3 choices of toppings, and 2 choices of meat. Multiply these choices: $2 \times 3 \times 2 = 12$ different ways.

Check your answer. To check this answer, divide the total number of outcomes by the number of choices in the first category. This becomes 12 ÷ 2 = 6. Next, divide this amount by the number of choices in the second category. This becomes 6 ÷ 3 = 2, which is the number of choices in the third category. This answer is checking.

PRACTICE 2

1. *Read and understand the question.* This question is looking for the total number of permutations, or orders, of 4 different objects taken 4 at a time.

Make a plan. Multiply the number of choices for each placement of the objects. An object can be used only once.

Carry out the plan. There are 4 choices for the first object, 3 for the second, 2 for the third, and 1 for the fourth. The number of permutations is therefore $4 \times 3 \times 2 \times 1 = 24$.

Check your answer. One way to check this solution is to divide the total number of permutations by each of the factors that were multiplied to see if the result is 1: $24 \div 4 = 6$, $6 \div 3 = 2$, and $2 \div 2 = 1$, so this solution is checking.

2. *Read and understand the question.* This question is looking for the total number of permutations, or orders, of 6 different trophies taken 6 at a time.

   Make a plan. Multiply the number of choices for each placement of the objects. An object can be used only once.

   Carry out the plan. There are 6 choices for the first object, 5 for the second, 4 for the third, 3 for the fourth, 2 for the fifth, and only 1 for the sixth. The number of permutations is therefore $6 \times 5 \times 4 \times 3 \times 2 \times 1 = 720$.

   Check your answer. One way to check this solution is to divide the total number of permutations by each of the factors that were multiplied to see if the result is 1.

   $$720 \div 6 = 120$$
   $$120 \div 5 = 24$$
   $$24 \div 4 = 6$$
   $$6 \div 3 = 2$$

   and

   $$2 \div 2 = 1$$

   so this solution is checking.

3. *Read and understand the question.* This question is looking for the total number of permutations, or orders, of 7 different objects taken 3 at a time.

   Make a plan. Multiply the number of choices for each placement of the objects until 3 factors are used. An object can be used only once.

   Carry out the plan. There are 7 choices for the first object, 6 for the second, and 5 for the third. The number of permutations is therefore $7 \times 6 \times 5 = 210$.

   Check your answer. One way to check this solution is to divide the total number of permutations by each of the factors that were multiplied to see if the result is 1.

   $$210 \div 7 = 30$$
   $$30 \div 6 = 5$$

   and

   $$5 \div 5 = 1$$

   so this solution is checking.

Practice 3

1. *Read and understand the question.* This question is looking for the total number of combinations of 2 books selected from a series of 4 books. *Make a plan.* Because the order is not important, find the number of permutations of 4 books taken 2 at a time, and then divide by the number of permutations of 2 books taken 2 at a time. *Carry out the plan.* The number of permutations of 4 students taken 2 at a time is equal to 4×3 and the total number of permutations of 2 books taken 2 at a time is equal to 2×1. Divide to find the total number of combinations of 4 books taken 2 at a time: $\frac{4 \times 3}{2 \times 1} = \frac{12}{2} = 6$ different combinations. *Check your answer.* To check this solution, reevaluate the number of combinations. The number of combinations is equal to $\frac{4 \times 3}{2 \times 1} = \frac{12}{2} = 6$, so this solution is checking.

2. *Read and understand the question.* This question is looking for the total number of combinations of 8 movies taken 4 at a time. *Make a plan.* Because the order is not important, find the number of permutations of 8 movies taken 4 at a time, and then divide by the number of permutations of 4 movies taken 4 at a time. *Carry out the plan.* The number of permutations of 8 movies taken 4 at a time is equal to $8 \times 7 \times 6 \times 5$ and the total number of permutations of 4 movies taken 4 at a time is equal to $4 \times 3 \times 2 \times 1$. Divide to find the total number of combinations of 8 movies taken 4 at a time:

$$\frac{8 \times 7 \times 6 \times 5}{4 \times 3 \times 2 \times 1} = \frac{1,680}{24} = 70$$

different combinations. *Check your answer.* To check this solution, reevaluate the number of combinations. The number of combinations is equal to

$$\frac{8 \times 7 \times 6 \times 5}{4 \times 3 \times 2 \times 1} = \frac{1,680}{24} = 70$$

so this solution is checking.

3. *Read and understand the question.* This question is looking for the total number of combinations of 6 students taken 4 at a time.

Make a plan. Because the order is not important, find the number of permutations of 6 students taken 4 at a time, and then divide by the number of permutations of 4 students taken 4 at a time.

Carry out the plan. The number of permutations of 6 students taken 4 at a time is equal to $6 \times 5 \times 4 \times 3$ and the total number of permutations of 4 students taken 4 at a time is equal to $4 \times 3 \times 2 \times 1$. Divide to find the total number of combinations of 6 students taken 4 at a time:

$$\frac{6 \times 5 \times 4 \times 3}{4 \times 3 \times 2 \times 1} = \frac{360}{24} = 15$$

different combinations.

Check your answer. To check this solution, reevaluate the number of combinations. The number of combinations is equal to

$$\frac{6 \times 5 \times 4 \times 3}{4 \times 3 \times 2 \times 1} = \frac{360}{24} = 15$$

so this solution is checking.

statistics word problems

Statistics means never having to say you're certain.
—Author Unknown

This lesson covers the measures of central tendencies such as mean, median, and mode. Word problems on these topics, as well as on range will also be explained.

THERE ARE THREE COMMON MEASURES of central tendencies. Each of these measures gives different types of information about the data in a set. The first type of measure is the mean.

MEAN

The mean is commonly known as the average of the numbers in a data set. To find the mean, find the sum of the set of numbers and then divide by the total number of values in the set. Take the following example.

The heights of players on a basketball team in inches are 69, 73, 75, 66, and 72. What is the mean height of the players on this team?

Read and understand the question. This question is looking for the mean of a set of 5 numbers.

Make a plan. Find the sum of the set of numbers and then divide by the number of values in the set. There are 5 values in the set, so divide the sum by 5.

Carry out the plan. The sum of the numbers is $69 + 73 + 75 + 66 + 72 = 355$. Divide the sum 355 by 5 to get 71. The average of the numbers is 71.

Check your answer. To check this solution, work backward and multiply the average by 5 to check to see if the total is 355: $71 \times 5 = 355$. This solution is checking.

What about the situation where the mean is given, but one or more of the data values are unknown? Use the strategy of working backward to find a missing value when the mean is known.

Charlie earned a 79, an 85, and an 88 on his first 3 history exams. What does he need to earn on his fourth test to have exactly an average of 85 for the 4 exams?

Read and understand the question. This question is looking for the grade on the fourth exam when the grades of the first 3 exams and the average of the 4 exams are given.

Make a plan. Find the total number of points needed on 4 tests to have an average of 85 by multiplying 85 by 4. Then, find the sum of the first 3 exams. Find the difference between these 2 amounts to find the score needed on the fourth exam.

Carry out the plan. The total needed on the four exams is $85 \times 4 = 340$. The sum total of the exams taken so far is $79 + 85 + 88 = 252$. The difference in these 2 amounts is $340 - 252 = 88$. He needs an 88 on the fourth exam.

Check your answer. To check this solution, find the average of the 4 exams to make sure it is equal to 85. The sum total of the 4 exams is $79 + 85 + 88 + 88 = 340$ and 340 divided by 4 is equal to 85. This solution is checking.

MEDIAN

The median of a data set is the number in the middle of the set when the values are placed in order. If there is an even number of values in the set, find the average of the two numbers in the middle of the set. Read through the following example.

The heights of players on a basketball team in inches are 69, 73, 75, 66, and 72. What is the median height?

Read and understand the question. This question is looking for the median value of a set of data.

Make a plan. The median of a set of data is the middle value in the list when the numbers are placed in order. Arrange the data values in order and find the middle value to find the median.

Carry out the plan. The numbers placed in order are 66, 69, 72, 73, 75. The middle value in this list is the third number, 72.

Check your answer. To check this solution, be sure that the numbers are placed in order and that no two values share the middle. There are five values, so the third value is the median. This solution of 72 is checking.

A new player with a height of 70 inches joins the team. What is the new median height of the team?

Read and understand the question. This question is looking for the median value of a set of data.

Make a plan. Re-arrange the data values in order and find the middle value to find the median. If two values share the middle, find the mean of these two values to find the median.

Carry out the plan. The numbers in order are 66, 69, 70, 72, 73, 75. The values 70 and 72 share the middle. The mean of 70 and 72 is equal to $\frac{70+72}{2} = \frac{142}{2} =$ 71. The median is 71.

Check your answer. To check the solution, make sure that the values are in order. Since two values share the middle, the mean of these values is the median. The mean of 70 and 72 is 71, so this solution is checking.

MODE

The mode of a set of data is the value that appears the most in the set.

The number of points scored by a basketball team in five games is 68, 72, 74, 66, and 72. What is the mode of the points scored?

Read and understand the question. This question is looking for the mode of a set of data.

Make a plan. The mode is the value that appears most often in a set of data. Count the number of times each value appears to find the number that appears most.

Carry out the plan. The number 72 appears in the list two times; each of the other values appears only once. The number 72 is the mode of this set of data.

Check your answer. To check this solution, double-check to be sure that no other value appears two or more times in the list. The number 72 is the only value that is repeated, so this solution is checking.

....

TIP: If two modes appear in a data set, the set is considered to be **bimodal.** If no value appears more than any other in the set, the set is considered to have **no mode.**

....

While mean, median, and mode are measures of central tendency, range is a measure of dispersion, or variability.

RANGE

The range of a set of data is the difference between the greatest value and the least value in the set.

The heights of players on a basketball team in inches are 68, 72, 74, 66, and 72. What is the range of the heights of the players?

Read and understand the question. This question is looking for the range of heights of five basketball players.

Make a plan. Find the difference between the tallest player and the shortest player to find the range.

Carry out the plan. The tallest player is 74 inches and the shortest is 66 inches. The difference is $74 - 66 = 8$. The range is 8 inches.

Check your answer. To check this solution, find the range again and be sure to subtract the greatest number from the least number in the set. The range for this set of values is $74 - 66 = 8$, so this solution is checking.

....

TIP: These four statistics can be summarized as the following:

The **mean** is the average value.
The **median** is the middle value after the data has been ordered.
The **mode** is the value the occurs the most often.
The **range** is the difference between the highest and lowest values.

....

PRACTICE

1. Jake earned a 95, 92, 88, and 89 on his latest science exams. What is his mean score for these 4 tests?

2. Using the information in question 1, Jake would like to have a 92 mean average in science class. If there will be a total of 5 exams, what score does he need on the next test to have a mean of 92 over the 5 exams?

3. The average daily temperatures in degrees over one week were 40, 44, 50, 38, 58, 42, and 39, respectively. What is the median temperature for the week?

4. The weights in pounds of 4 dogs at a kennel are 43, 65, 70, and 23. What is the median weight of these dogs?

5. The times in seconds for 6 swimmers in a certain swimming event are 50, 58, 59, 42, 43, and 58, respectively. What is the mode?

6. In a golf tournament, the scores of 5 different players were 74, 75, 75, 80, and 74. What is the mode of the set of golf scores?

7. The prices of 4 different shirts at a store are $10.99, $9.99, $14.99, and $19.99. What is the range in prices of these shirts?

ANSWERS

Practice

1. *Read and understand the question.* This question is looking for the mean of a set of 4 numbers.
   Make a plan. Find the sum of the set of numbers and then divide by the number of values in the set. There are 4 numbers in the set, so divide the sum by 4.
   Carry out the plan. The sum of the numbers is $95 + 92 + 88 + 89 = 364$. Divide the sum 364 by 4 to get 91. The mean of the numbers is 91.
   Check your answer. To check this solution, work backward and multiply the average by 4 to check to see if the total is 364: $91 \times 4 = 364$. This solution is checking.

2. *Read and understand the question.* This question is looking for the grade on the fifth exam when the grades of the first 4 exams and the mean of the 5 exams are given.

   Make a plan. Find the total number of points needed on 5 tests to have an average of 92 by multiplying 92 by 5. Then, find the sum of the first 4 exams. Find the difference between these two amounts to find the score needed on the fifth exam.

   Carry out the plan. The total needed on the 5 exams is $92 \times 5 = 460$. The sum of the exams taken so far is $95 + 92 + 88 + 89 = 364$. The difference in these two amounts is $460 - 364 = 96$. He needs a 96 on the fifth exam to have a mean average of 92.

   Check your answer. To check this solution, find the average of the 5 exams to make sure it is equal to 92. The sum of the 5 exams is $95 + 92 + 88 + 89 + 96 = 460$; 460 divided by 5 is equal to 92. This solution is checking.

3. *Read and understand the question.* This question is looking for the median value of a set of data.

   Make a plan. The median of a set of data is the middle value in the list when the numbers are placed in order. Arrange the data values in order and find the middle value to find the median.

   Carry out the plan. The numbers placed in order are 38, 39, 40, 42, 44, 50, 58. The middle value in this list is the fourth number, 42.

   Check your answer. To check this solution, be sure that the numbers are placed in order and that no two values share the middle. There are 7 values, so the fourth value is the median. This solution of 42 is checking.

4. *Read and understand the question.* This question is looking for the median value of a set of data.

   Make a plan. Arrange the data values in order, and find the middle value to find the median. If 2 values share the middle, find the mean of these 2 values to find the median.

   Carry out the plan. The numbers in order are 23, 43, 65, 70. The values 43 and 65 share the middle. The mean of 43 and 65 is equal to $\frac{43 + 65}{2} = \frac{108}{2} = 54$ The median is 54.

   Check your answer. To check the solution, make sure that the values are in order. Since two values share the middle, the mean of these values is the median. The mean of 43 and 65 is 54, so this solution is checking.

5. *Read and understand the question.* This question is looking for the mode of a set of data.

   Make a plan. The mode is the value that appears the most often in a set of data. Count the number of times each value appears to find the number that appears the most.

   Carry out the plan. The list is 50, 58, 59, 42, 43, and 58. The number 58 appears in the list two times; each of the other values appears only once. The number 58 is the mode of this set of data.

   Check your answer. To check this solution, double-check to be sure that no other value appears two or more times in the list. The number 58 is the only value that is repeated, so this solution is checking.

6. *Read and understand the question.* This question is looking for the mode of a set of data.

   Make a plan. The mode is the value that appears the most often in a set of data. Count the number of times each value appears to find the number that appears the most.

   Carry out the plan. The list is 74, 75, 75, 80, and 74. The numbers 74 and 75 each appear in the list two times; the other value appears only once. Therefore, the set has two modes and is called bimodal. The values 74 and 75 are the modes of this set of data.

   Check your answer. To check this solution, double-check to be sure that no other values appear two or more times in the list. The numbers 74 and 75 each appear twice, so this answer is checking.

7. *Read and understand the question.* This question is looking for the range of the prices of different shirts.

   Make a plan. Find the difference between the most expensive shirt and the least expensive shirt to find the range.

   Carry out the plan. The prices of the shirts are $10.99, $9.99, $14.99, and $19.99. The most expensive shirt is $19.99 and the least expensive shirt is $9.99. The difference is $19.99 – $9.99 = $10.00. The range is $10.

   Check your answer. To check this solution, find the range again and be sure to subtract the greatest number from the least number in the set. The range for this set of values is $19.99 – $9.99 = $10.00, so this answer is checking.

P O S T T E S T

THIS POSTTEST HAS 30 multiple-choice questions and is in the same format as the pretest. Each question corresponds with the concepts studied in the lesson in this book with the same number. Take the posttest to identify your areas of improvement. You may also use the posttest to help indicate areas where there is room for some growth in the skills and strategies needed for solving math word problems.

Read and answer each problem carefully by selecting the best answer choice. Don't be afraid to show your work; it is the cornerstone of problem solving and will also help you to identify any mistakes. When you have completed the posttest, check your responses with the answer key beginning on page 326. Each answer explanation uses the same format used for word-problem solving throughout the book. Use these steps to help create your plan for word-problem solving success.

POSTTEST

1. Which statement means the same as "four more than five times a number, *n*"?
   a. $5n + 4$
   b. $4n + 5$
   c. $n + 4$
   d. $4(5n)$

2. Sarevah's team is playing in a soccer tournament. During the tournament, each team will play each of the other teams exactly once. If there are 5 teams in the tournament, what is the total number of games that will be played during the tournament?
   a. 5 games
   b. 10 games
   c. 15 games
   d. 25 games

3. Teresa has 6 more pencils in her case than Ken. Becky has twice as many pencils as Teresa. If Ken has 12 pencils, how many does Becky have?
   a. 12 pencils
   b. 18 pencils
   c. 24 pencils
   d. 36 pencils

4. Kevin needs a ride and decides to take a taxicab. If it costs $6.50 to ride 4 miles in the cab, then what is the total amount it would cost to ride 16 miles at the same rate?
   a. $19.50
   b. $22.50
   c. $26.00
   d. $104.00

5. Sherry has a bank that contains only quarters and dimes. There is a total of $2.40 in the bank. What is the greatest number of quarters that she could have in her bank?
   a. 7 quarters
   b. 8 quarters
   c. 9 quarters
   d. 10 quarters

6. Todd is thinking of a number. His number is equal to two times the sum of 16 and 4. What is his number?

   a. 20

   b. 32

   c. 36

   d. 40

7. A DVD player is on sale for $\frac{2}{3}$ of the original price. If the original price was $69, what is the sale price of the player?

   a. $23

   b. $34.50

   c. $46

   d. $92

8. At a hot dog stand, Steve spent $18 on hot dogs for his family. If he bought a total of 8 hot dogs, how much did each one cost?

   a. $1.75

   b. $2.00

   c. $2.25

   d. $2.50

9. The sale price for 4 pounds of bananas is $1.32. What is the cost for 1 pound of bananas?

   a. $0.33

   b. $0.53

   c. $0.92

   d. $1.36

10. If 5 pounds of apples cost $8.45, what is the cost of 7 pounds?

   a. $1.69

   b. $8.52

   c. $9.15

   d. $11.83

11. Fifteen is what percent of 60?

   a. 4%

   b. 15%

   c. 20%

   d. 25%

12. On a science quiz, Peter received a 90%. If he answered 36 questions correctly and each question is worth the same number of points, what was the total number of questions on the test?
   a. 32 questions
   b. 38 questions
   c. 40 questions
   d. 46 questions

13. The product of 6 and a number increased by 10 is equal to 8 times the number. What is the number?
   a. 1.4
   b. 2
   c. 5
   d. 6

14. Joe wants to run at least 18 miles this week. If he runs 3 miles on Monday and twice as many miles on Wednesday, what is the minimum number of miles he still has to run this week to reach his goal?
   a. 6 miles
   b. 9 miles
   c. 12 miles
   d. 18 miles

15. The distance from Earth to the sun is approximately 150,000,000 kilometers (km). What is this distance in scientific notation?
   a. 1.5×10^7 km
   b. 15.0×10^8 km
   c. 1.5×10^8 km
   d. 15×10^8 km

16. Toby mixes coffee that costs $5 per pound with coffee that costs $7.50 per pound. He buys 4 more pounds of the coffee that costs $7.50 per pound than the coffee that costs $5 per pound. If he spends a total of $55 on coffee, how many pounds of coffee did he buy in all?
   a. 2 pounds
   b. 4 pounds
   c. 6 pounds
   d. 8 pounds

17. Two angles are supplementary. If the measure of one angle is 30 degrees less than the measure of the other angle, what is the measure of the smaller angle?

   a. 30 degrees

   b. 75 degrees

   c. 105 degrees

   d. 180 degrees

18. The measure of one angle of a triangle is 65 degrees, and the measure of the second angle is 10 degrees more than the first. What is the measure of the third angle of the triangle?

   a. 40 degrees

   b. 50 degrees

   c. 65 degrees

   d. 75 degrees

19. The measure of angle $\angle A$ in rhombus $ABCD$ is equal to twice the measure of angle B. What is the measure of angle $\angle A$?

   a. 30 degrees

   b. 60 degrees

   c. 90 degrees

   d. 120 degrees

20. The ratio of the corresponding sides of two similar triangles is 1:4. What is the ratio of their perimeters?

   a. 1:2

   b. 1:4

   c. 1:8

   d. 1:16

21. The measure of two different sides of an equilateral triangle can be expressed as $2x - 1$ and $4x - 11$. What is the measure of the perimeter of this triangle?

   a. 5

   b. 9

   c. 18

   d. 27

22. The circumference of a circle is $C = 16\pi$. How many units is the radius of this circle?

   a. 4 units

   b. 8 units

   c. 16 units

   d. 32 units

23. Richard is tiling a rectangular floor with a length of 12 feet and a width of 20 feet. If each tile will cover 4 square feet, how many tiles does he need?

   a. 20 tiles

   b. 24 tiles

   c. 60 tiles

   d. 240 tiles

24. A box in the shape of a cube has an edge of 3 meters. How many square meters of paper would be needed to cover the entire cube?

   a. $9 \, m^2$

   b. $27 \, m^2$

   c. $54 \, m^2$

   d. $81 \, m^2$

25. A toy box in the shape of a rectangular prism has a length of 4 feet, a width of 2 feet, and a height of 1.5 feet. What is the volume of the toy box?

   a. 12 cubic feet

   b. 15.5 cubic feet

   c. 24 cubic feet

   d. 34 cubic feet

26. Kevin is graphing a figure with the points $(-2,1)$, $(-3,4)$, and $(-8,8)$. In what quadrant is the figure located?

   a. Quadrant I

   b. Quadrant II

   c. Quadrant III

   d. Quadrant IV

27. Larry is playing a board game and needs to roll a 3 on a regular die on his next turn in order to win the game. What is the probability that Larry will win the game on his next turn?

a. $\frac{1}{6}$

b. $\frac{1}{3}$

c. $\frac{1}{2}$

d. 1

28. There are 6 black marbles and 3 red marbles in a bag. A marble is selected, not replaced, and then another is selected. What is the probability that both marbles selected are red?

a. 0

b. $\frac{1}{9}$

c. $\frac{1}{2}$

d. $\frac{1}{12}$

29. Ted has 5 different shirts and 6 pairs of pants in his closet. How many different outfits can he make by selecting one shirt and one pair of pants?

a. 6 outfits

b. 11 outfits

c. 30 outfits

d. 36 outfits

30. The elevations of 5 mountains are 29,035 feet, 28,169 feet, 27,765 feet, 27,920 feet, and 28,253. What is the median elevation of these mountains?

a. 1,270 feet

b. 28,169 feet

c. 28,228 feet

d. 28,253 feet

ANSWERS

1. **a.** *Read and understand the question.* This question is asking you to translate from words into math symbols.

   Make a plan. Translate using the key words and phrases in the question.

   Carry out the plan. The key phrase *more than* means addition. The phrase "five times a number" translates to $5n$. The final expression is $5n + 4$.

   Check your work. The only choices with *five times a number* written correctly are choices **a** and **d**. Choice **d** is not correct because 4 is being multiplied, not added.

2. **b.** *Read and understand the question.* This question is asking for the total number of games to be played in a tournament. Each of the 5 teams plays each team exactly once.

   Make a plan. Call the teams A, B, C, D, and E, and make an organized list of all of the games that will be played. Note that if Team A plays Team B, that is the same as Team B plays Team A.

   Carry out the plan. The organized list could look like the following:

A plays B	A plays C	A plays D	A plays E
	B plays C	B plays D	B plays E
		C plays D	C plays E
			D plays E

   This is a total of 10 games.

   Check your work. The number of games played can also be found by adding $4 + 3 + 2 + 1 = 10$, which is the same solution reached with the organized list.

3. **d.** *Read and understand the question.* This question is looking for the number of pencils Becky has. Information is also given about Ken's number of pencils and Teresa's number of pencils.

   Make a plan. Use the problem solving strategy of working backward, and start with the fact that Ken has 12 pencils.

   Carry out the plan. Since Ken has 12 pencils, and Teresa has 6 more than Ken, Teresa has $12 + 6 = 18$ pencils. Because Becky has twice as many as Teresa has, Becky has $2 \times 18 = 36$ pencils.

   Check your work. Start with 12, add 6 to get 18, and double this amount to get 36.

4. c. *Read and understand the question.* Kevin's 4-mile cab ride costs $6.50. Find the total cost of going 16 miles.

Make a plan. Look for a pattern using the fact that 4 miles costs $6.50, and go up 4 miles each time by adding another $6.50 until reaching 16 miles.

Carry out the plan. The pattern may look like the following:

Number of miles	4	8	12	16
Total cost	$6.50	$13.00	$19.50	$26.00

The total cost of 16 miles is $26.00.

Check your work. Another way to solve this problem is to set up a proportion and cross multiply. The proportion could be set up as $\frac{\$6.50}{4 \text{ miles}} = \frac{\$x}{16 \text{ miles}}$. Cross multiply to get $4x = 104$. Divide each side of the equation by 4.

$$\frac{4x}{4} = \frac{104}{4}$$
$$x = \$26$$

which is the same solution as the previous method.

5. b. *Read and understand the question.* This question is asking for the greatest number of quarters in a bank containing only quarters and dimes. The total amount of money in the bank is $2.40.

Make a plan. Use the strategy of guess and check. Remember to look for the greatest number of quarters. When using this strategy, do at least three trials to be sure of your solution.

Carry out the plan. Since 10 quarters is equal to $2.50, try 9 quarters first. Nine quarters is equal to $2.25. There is no way to add a number of dimes to this amount to get exactly $2.40. For the second trial, try 8 quarters. Eight quarters is equal to $2.00, so add 4 dimes to this amount to get $2.40. To be sure, try 7 quarters. Seven quarters is equal to $1.75, and there is no way to add a number of dimes to this amount and get exactly $2.40. The greatest number of quarters is 8.

Check your work. Eight quarters is equal to $2.00, added to 4 dimes is equal to $2.00 + $0.40 = $2.40.

6. d. *Read and understand the question.* This question asks for the number of which Todd is thinking, with a few clues given about the number.

Make a plan. Find a number that is equal to 2 times the sum of 16 and 4.

Carry out the plan. First, find the sum of 16 and 4. *Sum* is a key word for addition, so $16 + 4 = 20$. Then, find 2 times this sum: $2 \times 20 = 40$. His number is 40.

Check your work. $2(16 + 4) = 2(20) = 40$. This answer is checking.

7. c. *Read and understand the question.* This question is looking for the sale price of a DVD player when the original price and the fractional part of the discount are given.

Make a plan. To find the sale price, find $\frac{2}{3}$ of the original price by multiplying $\frac{2}{3} \times \$69$.

Carry out the plan.

$$\frac{2}{3} \times \$69 = \frac{2}{3} \times \frac{69}{1} = \frac{138}{3} = 46$$

The sale price is $46.

Check your work. If the sale price is $\frac{2}{3}$ of the original price, you are saving $\frac{1}{3}$ of the cost.

$$\frac{1}{3} \text{ of } 69 = \frac{1}{3} \times \frac{69}{1} = \frac{69}{3} = 23$$

You would save $23. To find the sale price, subtract the discount.

$$\$69 - \$23 = \$46$$

which is the same sale price calculated by the other method.

8. c. *Read and understand the question.* You are looking for the price of one hot dog when the cost of 8 is given.

Make a plan. To find the cost for one, divide the total amount of money spent by the number of hot dogs purchased.

Carry out the plan: $18 \div 8 = \$2.25$. The cost for each hot dog is $2.25.

Check your work. To check this problem, multiply the price of one hot dog by 8: $\$2.25 \times 8 = \$18.00$, which is the total amount that Steve spent.

9. a. *Read and understand the question.* This question is looking for the unit rate for 1 pound of bananas when the price of 4 pounds is given.

Make a plan. Divide the cost of 4 pounds by 4 to find the price for 1 pound.

Carry out the plan: $\$1.32 \div 4 = \$0.33$. One pound of bananas is $0.33.

Check your work. To check, multiply $0.33 by 4 to find the cost of 4 pounds. This is equal to $1.32, so this answer is checking.

10. d. *Read and understand the question.* This question is looking for the total cost of 7 pounds of apples when the cost of 5 pounds is given.

Make a plan. Set up a proportion comparing the cost with the number of pounds. The proportion could be set up as follows:

$$\frac{\text{cost}(\$)}{\text{number of pounds}} = \frac{\text{cost}(\$)}{\text{number of pounds}}$$

Carry out the plan. Use the given values in the proportion and cross multiply.

$$\frac{\$8.45}{5 \text{ pounds}} = \frac{\$x}{7 \text{ pounds}}$$

Cross multiply: $5x = 59.15$. Divide each side of the equation by 5.

$$\frac{5x}{5} = \frac{59.15}{5}$$

$$x = 11.83$$

The cost of 7 pounds is $11.83.

Check your work. Since the price of 5 pounds is $8.45, then the price for one pound is equal to $8.45 divided by 5. The unit price is $1.69. Multiply $1.69 by 7 to get the cost for 7 pounds. This is also equal to $11.83, so this answer is checking.

11. d. *Read and understand the question.* You are asked to find the percent that 15 is of 60.

Make a plan. Set up the proportion $\frac{\text{part}}{\text{whole}} = \frac{\%}{100}$, where 15 is the part, 60 is the whole, and x is the percent to be found.

Carry out the plan. The proportion becomes $\frac{15}{60} = \frac{x}{100}$. Cross multiply to get $60x = 1{,}500$. Divide each side of the equation by 60.

$$\frac{60x}{60} = \frac{1500}{60}$$

$$x = 25$$

15 is 25% of 60.

Check your work. To check this problem, find 25% of 60 by multiplying 0.25×60. This is equal to 15, so this answer is checking.

12. c. *Read and understand the question.* This problem is asking for the number of questions on a test when the percent answered correctly and the number of questions answered correctly is known.

Make a plan. Set up a proportion comparing the correct number of questions to the corresponding percent earned, over the total of 100%.

Carry out the plan. The proportion could be set up as

$$\frac{\text{number of questions correct}}{\text{total number of questions}} = \frac{\text{percent of questions correct}}{100} = \frac{36}{x} = \frac{90}{100}$$

Cross multiply to get $90x = 3{,}600$. Divide each side of the equation by 90.

$$\frac{90x}{90} = \frac{3{,}600}{90}$$

$$x = 40$$

There are 40 questions on the test.

Check your work. Set up the proportion using 40 as the total number of questions. Then, cross multiply to be sure the cross products are equal.

$$\frac{36}{40} = \frac{90}{100}$$

$$3{,}600 = 3{,}600$$

so the proportion is checking.

13. c. *Read and understand the question.* This question is looking for a number when clues about the number are given.

Make a plan. Use the key words and phrases in the problem to write an equation using mathematical symbols. Then, solve the equation to find the number.

Carry out the plan. Let n = the number. The first part of the statement translates to $6n + 10$ and the second part translates to $8n$. Set these parts equal and solve the equation $6n + 10 = 8n$. Subtract $6n$ from each side of the equation.

$$6n - 6n + 10 = 8n - 6n$$

The equation becomes $10 = 2n$. Divide each side of the equation by 2 to get $n = 5$. The number is 5.

Check your work. Substitute 5 for "the number" in the question. The product of six and five is 30, increased by 10 is $30 + 10 = 40$. Eight times 5 is equal to 40. The results are equal, and the answer is checking.

14. b. *Read and understand the question.* This question is looking for the minimum number of miles Joe should run when he has already run some of the miles so far this week.

Make a plan. Write an inequality that relates the miles he has run so far to the goal of 18 miles.

Carry out the plan. Let m = the minimum number of miles he still needs to run. He ran 3 miles on Monday, plus twice as many (6) on Tuesday. Add these amounts to m and set the sum greater than or equal to 18. The inequality is $3 + 6 + m \geq 18$. Combine like terms: $9 + m \geq 18$. Subtract 9 from each side: $9 - 9 + m \geq 18 - 9$. The inequality is $m \geq 9$. Joe must run a minimum of 9 miles.

Check your work. Add the amounts for Monday, Tuesday, and the miles to go together: $3 + 6 + 9 = 18$, which was the minimum amount. This answer is checking.

15. c. *Read and understand the question.* This question is asking for a number to be changed to scientific notation from standard form.

Make a plan. Write the non-zero numbers as a value between 1 and 10. Multiply this value by a power of 10, where the exponent is the number of places the decimal moves to the left. The exponent is positive since the original number is greater than 1.

Carry out the plan. Place the decimal point between 1 and 5 to create a number between 1 and 10. The exponent is 8 because the decimal point has been moved 8 places to the left. The scientific notation is 1.5×10^8.

Check your work. To check this problem, take 1.5 and move the decimal point 8 places to the right. Add zeros where needed. This makes the number 150,000,000, which was the original number. This answer is checking.

16. d. *Read and understand the question.* You are looking for the total number of pounds of coffee that is bought when two different types are purchased. The price per pound and the total amount spent are given.

Make a plan. There are 4 more pounds of the more expensive coffee, so let x represent the less expensive coffee and $x + 4$ represent the more expensive coffee. Write an expression for the sum of the money spent on both types and set it equal to $55.

Carry out the plan. Let $x =$ the number of pounds of the $5.00 coffee, and let $x + 4 =$ the number of pounds of the $7.50 coffee. Write the equation

$$\$5.00(x) + \$7.50(x + 4) = \$55.00$$

Use the distributive property within the equation to get $5x + 7.5x + 30 = 55$. Combine like terms to get $12.5x + 30 = 55$. Subtract 30 from each side of the equation: $12.5x + 30 - 30 = 55 - 30$. Simplify: $12.5x = 25$. Divide each side of the equation by 12.5:

$$\frac{12.5x}{12.5} = \frac{25}{12.5}$$

Thus, $x = 2$ and $x + 4 = 2 + 4 = 6$. Toby bought 2 pounds of the $5.00 coffee and 6 pounds of the $7.50 coffee. This is a total of $2 + 6 = 8$ pounds.

Check your work. Two pounds of the $5.00 per pound coffee costs $2(\$5) = \$10$. Six pounds of the $7.50 per pound coffee costs $6(\$7.50) = \$45$. The total for both types is $\$10 + \$45 = \$55$, which was the total amount spent in the problem. This answer is checking.

17. b. *Read and understand the question.* This question is looking for the measure of the smaller of two supplementary angles.

Make a plan. The sum of two supplementary angles is 180 degrees. Use the information given about the two angles to write an equation set equal to 180.

Carry out the plan. Let $x =$ the larger angle and let $x - 30 =$ the smaller angle. The term *supplementary* means that the sum of the two angles is 180 degrees. To write the equation, add the two angles and set the sum equal to 180.

$$x + x - 30 = 180$$

Combine like terms to get $2x - 30 = 180$. Add 30 to each side of the equation.

$$2x - 30 + 30 = 180 + 30$$

The equation simplifies to $2x = 210$. Divide each side of the equation by 2.

$$\frac{2x}{2} = \frac{210}{2}$$
$$x = 105$$

Therefore, the larger angle is 105 degrees and the smaller angle is $x - 30 = 105 - 30 = 75$ degrees.

Check your work. To check this problem, add the two angle measures: $105 + 75 = 180$. Because the two angles are supplementary, they add to 180 degrees. This answer is checking.

18. a. *Read and understand the question.* This question is asking for the measure of the third angle of a triangle when the measure of the first and a clue about the second angle is given.

Make a plan. The sum of the three angles of a triangle is 180 degrees. Use the measure of the first angle to find the measure of the second. Then, subtract the sum of these two angles from 180.

Carry out the plan. The first angle measures 65 degrees. Because the second angle is 10 degrees more than the first, the second angle measures $65 + 10 = 75$ degrees. Subtract the sum of these two angles from 180: $180 - (65 + 75) = 180 - 140 = 40$ degrees. The third angle is 40 degrees.

Check your work. Add the measures of all three angles to be sure that they add to 180 degrees: $65 + 75 + 40 = 180$. This answer is checking.

19. d. *Read and understand the question.* You need to find the measure of angle $\angle A$ in rhombus $ABCD$ when information is given about two consecutive angles. The sum of any two consecutive angles in a rhombus is 180 degrees.

Make a plan. The measure of angle $\angle A$ is twice the measure of angle $\angle B$. Write an expression for the sum of these two angles, and set the sum equal to 180 degrees to solve for the angle measures.

Carry out the plan. Angle $\angle A$ and angle $\angle B$ in rhombus $ABCD$ are consecutive angles, or angles that are next to each other. Therefore, the sum of their measures is 180 degrees. Let $x =$ the measure of angle $\angle B$, then $2x =$ the measure of angle $\angle A$. Because the sum of the two angles is 180 degrees, write the equation $x + 2x = 180$. Combine like terms: $3x = 180$. Divide each side of the equation by 3.

$$\frac{3x}{3} = \frac{180}{3}$$
$$x = 60$$

Thus, the measure of angle $\angle B$ is 60 degrees, and the measure of angle $\angle A$ is $2(60) = 120$ degrees.

Check your work. Add the measures of the consecutive angles $\angle A$ and $\angle B$ to be sure that the sum is 180 degrees: $60 + 120 = 180$. This answer is checking.

20. b. *Read and understand the question.* This question asks for the ratio of the perimeters of two similar triangles when the ratio of the corresponding sides is given.

Make a plan. Use the strategy of guess and check to try values for the sides of the triangles in the ratio of 1:4. Then, find the perimeters and compare their ratio.

Carry out the plan. Use the sides 2, 3, and 4 for the smaller triangle. Since the ratio of corresponding sides is 1:4, multiply each of these values by 4 to find the sides of the larger triangle. Therefore, the sides would be 8, 12, and 16. Now, add the sides of each triangle to find the perimeters. The smaller triangle has a perimeter of $2 + 3 + 4 = 9$, and the larger triangle has a perimeter of $8 + 12 + 16 = 36$. Because the ratio of 9:36 simplifies to 1:4, the ratio of their perimeters is 1:4. This is the same as the ratio of the corresponding sides.

Check your work. To check this solution, divide each of the perimeters by 9: $9 \div 9 = 1$ and $36 \div 9 = 4$. This is the ratio 1:4. The answer is checking.

21. d. *Read and understand the question.* Two expressions are given for two sides of an equilateral triangle. Use this information to find the perimeter of the triangle.

Make a plan. Because all sides of an equilateral triangle are the same measure, set the two given expressions equal to each other and solve for x. Use this value of x to substitute and find the perimeter.

Carry out the plan. Set the two known expressions equal to each other: $2x - 1 = 4x - 11$. Subtract $2x$ from each side of the equation:

$$2x - 2x - 1 = 4x - 2x - 11$$

The equation simplifies to $-1 = 2x - 11$. Add 11 to each side to get $10 = 2x$. Divide each side of the equation by 2 to get $x = 5$. Substitute this value into one of the expressions to find the length of one side: $2x - 1 = 2(5) - 1 = 10 - 1 = 9$. If one side is 9, then the perimeter is $3 \times 9 = 27$.

Check your work. Substitute the value of x into the other expression to be sure each side is 9.

$$4x - 11 = 4(5) - 11 = 20 - 11 = 9$$

The answer is checking.

22. b. *Read and understand the question.* In this question, you are asked to find the radius of a circle when the circumference is given.

Make a plan. Use the formula for circumference and the strategy of working backward to find the radius. The formula for the circumference of a circle is $C = \pi d$, where d is the diameter of the circle.

Carry out the plan. Because $C = \pi d$ and you are given $C = 16\pi$, then the diameter (d) is equal to 16. The radius of a circle is equal to half of the diameter, and half of 16 is 8. The radius is 8 units.

Check your work. Work forward through the problem to check. If the radius is 8 units, then the diameter is $8 \times 2 = 16$. Thus, the circumference of this circle is $C = \pi d = 16\pi$. This solution is checking.

23. c. *Read and understand the question.* You are asked to find the number of tiles needed to cover a rectangular area. The dimensions of the floor are given, and the size of the tiles is also known.

Make a plan. Find the area of the floor by using the formula *Area = base* $\times$ *height.* Then, divide this area by the area covered by each tile.

Carry out the plan. *Area* $= 12 \times 20 = 240$ square feet. Divide this area by 4 to find the number of tiles needed: $240 \div 4 = 60$. He will need 60 tiles.

Check your work. Work backward to check this problem: 60 tiles will cover $60 \times 4 = 240$ square feet. Because the area of the rectangular floor is also $12 \times 20 = 240$ square feet, this is the correct number of tiles. This solution is checking.

24. c. *Read and understand the question.* This question asks for the amount of paper needed to cover a cube; in other words, the surface area of the cube.

Make a plan. Use the formula for surface area of a cube ($SA = 6e^2$, where e is the length of an edge of the cube). Substitute the length of the edge (3 meters) given into the formula.

Carry out the plan. The formula becomes $SA = 6(3)^2$. Evaluate the exponent first: $SA = 6(9)$. Multiply: $SA = 54$ m^2.

Check your work. Work backward to check the solution. Begin with the surface area of 54 square meters. Divide this number by 6 faces: $54 \div 6 = 9$. This is the area of each face of the cube. Because $3 \times 3 = 9$, each edge measures 3 meters. This solution is checking.

25. a. *Read and understand the question.* This question asks for the volume of a rectangular prism when the dimensions are given.

Make a plan. Use the volume formula $V = l \times w \times h$, and substitute the values given in the question.

Carry out the plan. The volume formula becomes $V = 4 \times 2 \times 1.5$, so $V = 12$. The volume is 12 cubic units.

Check your work. Double-check your work by multiplying the values to be sure the correct volume is 12 cubic units: $4 \times 2 \times 1.5 = 12$. The answer is checking.

26. b. *Read and understand the question.* You are asked to find the quadrant in which a figure is plotted. There are four quadrants in the coordinate plane.

Make a plan. Examine the coordinates for each point, and graph each point starting at the origin. If the first number (x-coordinate) is positive, go to the right, and if the number is negative, count to the left. If the second number (y-coordinate) is positive, count up, and if it is negative, count down.

Carry out the plan. Each of the points has a negative x-coordinate and a positive y-coordinate. Each of the points should be graphed by counting over to the left and then up. The figure is a triangle located in Quadrant II, as shown in the following figure.

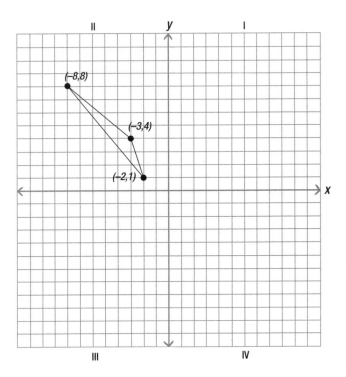

Check your work. Each of the points is located in the second quadrant, where x-coordinates are negative and y-coordinates are positive. This solution is checking.

27. a. *Read and understand the question.* This question is asking for the probability that Larry will roll a 3 on his next turn and will win the game.
Make a plan. The probability of an event *e* is equal to

$$P(E) = \frac{\text{the number of ways the event can occur}}{\text{total number of possible outcomes}}$$

There are 6 sides to the die, so there are 6 total outcomes. There is only one way to roll a 3.
Carry out the plan. The probability is $P(E) = \frac{1}{6}$.
Check your work. There is only one way to roll a 3 on a 6-sided die. This makes the probability equal to $\frac{1}{6}$. The answer is checking.

28. d. *Read and understand the question.* This question is looking for the probability of selecting a red marble first, and another red marble second. The events are dependent events because the first marble is not replaced after it is selected.
Make a plan. Use the fact that there are 3 red marbles and that there are a total of 9 marbles. Because the first marble is not replaced after it is chosen, the number of marbles decreases by one for the second draw. Find the probability of each event, and multiply them together.
Carry out the plan. The probability of first selecting a red marble at random is $\frac{3}{9} = \frac{1}{3}$, which leaves only 8 marbles left. The probability of selecting another red marble second is $\frac{2}{8} = \frac{1}{4}$. Multiply these probabilities to find the probability that both events will happen: $\frac{1}{3} \times \frac{1}{4} = \frac{1}{12}$
Check your answer. To check this solution, multiply the probabilities together again. Because $\frac{1}{3} \times \frac{1}{4} = \frac{1}{12}$, this answer is checking.

29. c. *Read and understand the question.* You need to find the total number of outfits that can be made from 5 shirts and 6 pairs of pants.
Make a plan. One way to solve this problem is to multiply the number of choices for each to find the total number of outfits.
Carry out the plan. Multiply 5 shirts by 6 pairs of pants: $5 \times 6 = 30$ different outfits.
Check your work. You can check your work by making an organized list of each shirt with each of the different pairs of pants. There are 30 different outfits in the list. This answer is checking.

30. b. *Read and understand the question.* Find the median elevation of a list of the tallest five mountains in the world.

Make a plan. The median elevation will be the value in the middle of the list, when the list is in order from smallest to largest. Put the values in order, and locate the middle value.

Carry out the plan. The order of the given elevations from smallest to largest is 27,765; 27,920; 28,169; 28,253, 29,035. The value in the middle of the list is 28,169 feet.

Check your work. You could also find the solution by listing the elevations in order from largest to smallest: 28,169 is the middle number in this list also. The answer is checking.

GLOSSARY

absolute value   the distance a number or expression is from zero on a number line

acute angle   an angle that measures less than 90 degrees

acute triangle   a triangle in which every angle measures less than 90 degrees

addend   any number to be added

adjacent angles   two angles that share a vertex and a common ray

angle   two rays connected by a vertex

arc   a curved section of a circle

area   a measure of how many square units it takes to cover a closed figure

associative law of addition   the property of numbers that allows you to regroup numbers when adding, that is, $a + (b + c) = (a + b) + c$

base   a number used as a repeated factor in an exponential expression

bimodal   a set of data that has exactly two modes

chord   a line segment that goes through a circle with its endpoints on the circle

circle   the set of all points equidistant from one given point, called the *center*. The center point defines the circle, but it is not on the circle.

circumference   the distance around a circle

coefficient   the number placed next to a variable in a term

combination   a group of objects in which the order does not matter

complementary angles   two angles whose sum is 90 degrees

compound inequality   a combination of two or more inequalities

commutative property of addition   the property of numbers that states that order does not matter when adding, that is, $a + b = b + a$.

congruent   identical in shape and size

consecutive even integers   even integers in order, such as 2, 4, 6, or –12, –10, –8

consecutive integers   numbers from the set of integers in order, such as 3, 4, 5, or –8, –7, –6

consecutive odd integers   odd integers in order, such as 1, 3, 5, or –11, -9, –7

continued ratio   a ratio comparison of three or more values

coordinate plane   a grid divided into four quadrants by both a horizontal x-axis and a vertical y-axis

cross product   a product of the numerator of one fraction and the denominator of a second fraction

decimal   a number related to or based on the number 10. The place value system is a decimal system because the place values (units, tens, hundreds, etc.) are based on 10.

denominator   the bottom number in a fraction. *Example:* 2 is the denominator in $\frac{1}{2}$.

dependent events   the events in a probability situation where the likeliness of the second event occurring depends on the outcome of the first event

diameter   a line segment that passes through the center of the circle whose endpoints are on the circle

difference   the difference between two numbers means subtract one number from the other

divisible by   a number is divisible by a second number if that second number divides *evenly* into the original number with no remainder. *Example:* 10 is divisible by 5 ($10 \div 5 = 2$, with no remainder). However, 10 is not divisible by 3.

distributive property   when multiplying a sum or a difference by a third number, you can multiply each of the first two numbers by the third number and then add or subtract the products, that is, $(a + b)c = ac + bc$

dividend   a number that is divided by another number

divisor   a number that divides another number

equation   a mathematical statement that can use numbers, variables, or a combination of the two and an equal sign

equilateral triangle   a triangle with three equal sides and three equal angles

even integer   integers that are divisible by 2, such as –4, –2, 0, 2, 4, and so on

exponent   a value that tells you how many times a number, the base, is a factor in the product

expression   a mathematical statement without an equal sign that can use numbers, variables, or a combination of the two

factor   a number that is multiplied to find a product

fraction   the result of dividing two numbers. When you divide 3 by 5, you get $\frac{3}{5}$, which equals 0.6. A fraction is a way of expressing a number that involves dividing a top number (the numerator) by a bottom number (the denominator).

greatest common factor   the largest of all the common factors of two or more numbers

hypotenuse   the longest leg of a right triangle, which is always opposite the right angle

improper fraction   a fraction in which the absolute value of the numerator is greater than or equal to the absolute value of the denominator

independent events   the events in a probability situation where the likeliness of the second event occurring does not depend on the outcome of the first event

inequality   a sentence that compares quantities that are greater than, less than, greater than or equal to, or less than or equal to each other

integer   a number along the number line, such as –3, –2, –1, 0, 1, 2, 3, and so on. Integers include whole numbers and their negatives.

inverse operation   the opposite of a given operation. *Example:* the inverse operation of addition is subtraction.

isosceles triangle   a triangle with two equal sides and two equal angles

least common denominator   the smallest number divisible by two or more denominators

least common multiple   the smallest of all the common multiples of two or more numbers

like terms   two or more terms that have exactly the same variable, with the variable raised to the same exponent

line   a straight path that continues forever in two directions

linear pair   two adjacent angles that together form a straight angle

major arc   an arc greater than or equal to 180 degrees

mean   the average of a set of data

median   the middle value in a set of numbers that are arranged in increasing or decreasing order. If there are two middle numbers, it is the average of these two.

midpoint the point at the exact middle of a line segment

minor arc an arc less than or equal to 180 degrees

mixed number a number with an integer part and a fractional part. Mixed numbers can be converted into improper fractions.

mode the value in a set of numbers that occurs most often. There can be one mode, several modes, or no mode.

multiple of a number is a multiple of a second number if that second number can be multiplied by an integer to get the original number. *Example:* 10 is a multiple of 5 $(10 = 5 \times 2)$; however, 10 is not a multiple of 3.

mutually exclusive events two or more events that do not occur simultaneously

natural numbers the set of numbers $\{1, 2, 3, 4, 5, \ldots\}$

negative number a number that is less than zero, such as $-1, -18.6, -14$.

numerator the top part of a fraction. *Example:* 1 is the numerator in $\frac{1}{2}$.

obtuse angle an angle that measures greater than 90 degrees but less than 180 degrees

obtuse triangle a triangle with an angle that measures greater than 90 degrees

odd integer integers that are not divisible by 2, such as $-5, -3, -1, 1, 3$, and so on

order of operations the order in which operations are performed

ordered pair a location of a point on a coordinate plane in the form (x,y)

origin coordinate pair $(0,0)$ and the point where the x- and y-axes intersect

parallel lines two lines that do not intersect

parallelogram a quadrilateral with two pairs of parallel sides

percent a ratio that compares numerical data to the hundred. The symbol for percent is %.

perimeter the distance around a figure

permutation a group of objects in which the order of the objects matters; when the order is changed, a different permutation occurs

perpendicular lines lines that intersect to form right angles

positive number a number that is greater than zero, such as $2, 42, \frac{1}{2}, 4.63$

prime factorization the process of breaking down factors into prime numbers

prime number an integer that is divisible only by 1 and itself, such as 2, 3, 5, 7, 11, and so on. All prime numbers are odd, except for 2. The number 1 is not considered prime.

probability the likelihood that a specific event will occur

product the answer to a multiplication problem

proper fraction a fraction in which the absolute value of the numerator is less than the absolute value of the denominator

proportion an equation that states that two ratios are equal

Pythagorean theorem   the formula $a^2 + b^2 = c^2$, where a and b represent the lengths of the legs and c represents the length of the hypotenuse of a right triangle

Pythagorean triple   a set of three integers that satisfies the Pythagorean theorem, such as 3:4:5

quadrilateral   a polygon with four sides

quotient   the answer you get when you divide. *Example:* 10 divided by 5 is 2; the quotient is 2.

radical   the symbol used to signify a root operation

radicand   the number inside a radical

radius   the line segment whose one endpoint is at the center of the circle and whose other endpoint is on the circle

range   the difference between the largest value and the smallest value in a set of data

ratio   a comparison of two values using numbers

rational numbers   the set of numbers that can be expressed as $\frac{a}{b}$, where $b \neq 0$ and a and b are integers

ray   a line that has one endpoint and continues forever in one direction

reciprocal   the multiplicative inverse of a fraction. *Example:* $\frac{2}{1}$ is the reciprocal of $\frac{1}{2}$.

rectangle   a parallelogram with four right angles

remainder   The number left over after division. *Example:* 11 divided by 2 is 5, with a remainder of 1.

rhombus   a parallelogram with four equal sides

right angle   an angle that measures exactly 90 degrees

right triangle   a triangle with an angle that measures exactly 90 degrees

scalene triangle   a triangle with no equal sides

simplify   to combine like terms and reduce an equation to its most basic form

slope   the steepness of a line, found by dividing the difference between two y-values by the difference between their corresponding x-values

square   a parallelogram with four equal sides and four right angles

square of a number   the product of a number and itself, such as 4^2, which is 4×4

sum   the sum of two numbers means the two numbers are added together

supplementary angles   two angles whose sum is 180 degrees

surface area   the sum of the area of the faces of a three-dimensional figure

table   graphic organizer that arranges information into columns and rows

term   a number or a number and the variables associated with it

triangle   a polygon with three sides

variables   letters used to stand in for numbers

vertex   a point at which two lines, line segments, or rays connect

vertical angles   two opposite congruent angles formed by intersecting lines

volume   a cubic measurement that measures how many cubic units it takes to fill a solid figure

whole number   any of the set of nonnegative integers {0, 1, 2, 3, 4, 5, . . .}

NOTES

NOTES

NOTES